# Python Debugging for AI, Machine Learning, and Cloud Computing

## A Pattern-Oriented Approach

**Dmitry Vostokov**

Apress®

*Python Debugging for AI, Machine Learning, and Cloud Computing:*
*A Pattern-Oriented Approach*

Dmitry Vostokov
Dalkey, Dublin, Ireland

ISBN-13 (pbk): 978-1-4842-9744-5    ISBN-13 (electronic): 978-1-4842-9745-2
https://doi.org/10.1007/978-1-4842-9745-2

Managing Director, Apress Media LLC: Welmoed Spahr
Acquisitions Editor: Celestin Suresh John
Development Editor: James Markham
Editorial Assistant: Gryffin Winkler
Copy Editor: Mary Behr

Cover designed by eStudioCalamar

Cover image designed by Igor Mamaev on pixabay

Distributed to the book trade worldwide by Springer Science+Business Media New York, 1 New York Plaza, Suite 4600, New York, NY 10004-1562, USA. Phone 1-800-SPRINGER, fax (201) 348-4505, e-mail orders-ny@springer-sbm.com, or visit www.springeronline.com. Apress Media, LLC is a California LLC and the sole member (owner) is Springer Science + Business Media Finance Inc (SSBM Finance Inc). SSBM Finance Inc is a **Delaware** corporation.

For information on translations, please e-mail booktranslations@springernature.com; for reprint, paperback, or audio rights, please e-mail bookpermissions@springernature.com.

Apress titles may be purchased in bulk for academic, corporate, or promotional use. eBook versions and licenses are also available for most titles. For more information, reference our Print and eBook Bulk Sales web page at www.apress.com/bulk-sales.

Any source code or other supplementary material referenced by the author in this book is available to readers on GitHub. For more detailed information, please visit https://www.apress.com/gp/services/source-code.

Paper in this product is recyclable

*To Ekaterina, Alexandra, Kirill, and Maria*

# Table of Contents

# About the Author

**Dmitry Vostokov** is an internationally recognized expert, speaker, educator, scientist, inventor, and author. He founded the pattern-oriented software diagnostics, forensics, and prognostics discipline (Systematic Software Diagnostics) and Software Diagnostics Institute (DA+TA: DumpAnalysis.org + TraceAnalysis.org). Vostokov has also authored multiple books on software diagnostics, anomaly detection and analysis, software, and memory forensics, root cause analysis and problem-solving, memory dump analysis, debugging, software trace and log analysis, reverse engineering, and malware analysis. He has over thirty years of experience in software architecture, design, development, and maintenance in various industries, including leadership, technical, and people management roles. In his spare time, he presents multiple topics on Debugging.TV and explores software narratology and its further development as narratology of things and diagnostics of things (DoT), software pathology, and quantum software diagnostics. His current interest areas are theoretical software diagnostics and its mathematical and computer science foundations, application of formal logic, artificial intelligence, machine learning, and data mining to diagnostics and anomaly detection, software diagnostics engineering and diagnostics-driven development, diagnostics workflow, and interaction. Recent interest areas also include cloud native computing, security, automation, functional programming, applications of category theory to software development and big data, and artificial intelligence diagnostics.

# About the Technical Reviewer

 **Krishnendu Dasgupta** is currently the Head of Machine Learning at Mondosano GmbH, leading data science initiatives focused on clinical trial recommendations and advanced patient health profiling through disease and drug data. Prior to this role, he co-founded DOCONVID AI, a startup that leveraged applied AI and medical imaging to detect lung abnormalities and neurological disorders.

With a strong background in computer science engineering, Krishnendu has more than a decade of experience in developing solutions and platforms using applied machine learning. His professional trajectory includes key positions at prestigious organizations such as NTT DATA, PwC, and Thoucentric.

Krishnendu's primary research interests include applied AI for graph machine learning, medical imaging, and decentralized privacy-preserving machine learning in healthcare. He also had the opportunity to participate in the esteemed Entrepreneurship and Innovation Bootcamp at the Massachusetts Institute of Technology, cohort of 2018 batch.

Beyond his professional endeavors, Krishnendu actively dedicates his time to research, collaborating with various research NGOs and universities worldwide. His focus is on applied AI and ML.

# Introduction

Python is the dominant language used in AI and machine learning with data and pipelines in cloud environments. Besides debugging Python code in popular IDEs, notebooks, and command-line debuggers, this book also includes coverage of native OS interfacing (Windows and Linux) necessary to understand, diagnose, and debug complex software issues.

The book begins with an introduction to pattern-oriented software diagnostics and debugging processes that, before doing Python debugging, diagnose problems in various software artifacts such as memory dumps, traces, and logs. Next, it teaches various debugging patterns using Python case studies that model abnormal software behavior. Further, it covers Python debugging specifics in cloud native and machine learning environments. It concludes with how recent advances in AI/ML can help in Python debugging. The book also goes deep for case studies when there are environmental problems, crashes, hangs, resource spikes, leaks, and performance degradation. It includes tracing and logging besides memory dumps and their analysis using native WinDbg and GDB debuggers.

This book is for those who wish to understand how Python debugging is and can be used to develop robust and reliable AI, machine learning, and cloud computing software. It uses a novel pattern-oriented approach to diagnosing and debugging abnormal software structure and behavior. Software developers, AI/ML engineers, researchers, data engineers, MLOps, DevOps, and anyone who uses Python will benefit from this book.

Source Code: All source code used in this book can be downloaded from `github.com/Apress/Python-Debugging-for-AI-Machine-Learning-and-Cloud-Computing`.

# CHAPTER 1

# Fundamental Vocabulary

Debugging complex software issues in machine learning and cloud computing environments requires not only the knowledge of the Python language and its interpreter (or compiler), plus standard and external libraries, but also necessary and relevant execution environment and operating system internals. In this chapter, you will review some necessary fundamentals from software diagnostics and debugging languages to have the same base level of understanding for the following chapters. In this book, I assume that you are familiar with the Python language and its runtime environment.

## Process

A Python script is interpreted by compiling it into bytecode and then executing it, or it can even be precompiled into an application program. In both cases, this interpreter file or the compiled application is an executable program (in Windows, it may have a `.exe` extension) that references some operating system libraries (`.dll` in Windows and .so in Linux). This application can be loaded into computer memory several times; each time, a separate process is created with its own resources and unique process ID (PID, also TGID), as shown in Figure 1-1. The process may also have a parent process that created it, with a parent process ID (PPID).

1

© Dmitry Vostokov 2024
D. Vostokov, *Python Debugging for AI, Machine Learning, and Cloud Computing*,
https://doi.org/10.1007/978-1-4842-9745-2_1

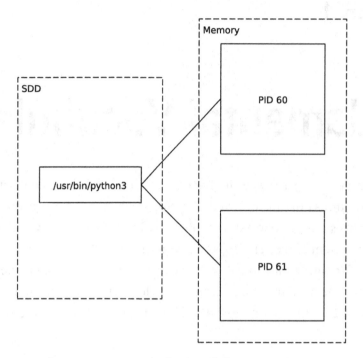

***Figure 1-1.*** *Two python3 processes with two different PIDs*

To illustrate, I executed the code in Listing 1-1 on both Windows and Linux twice.

***Listing 1-1.*** A Simple Script to Model Running Python Code

```python
import time

def main():
    foo()

def foo():
    bar()

def bar():
    while True:
        time.sleep(1)

if __name__ == "__main__":
    main()
```

Figure 1-2 shows two processes on Windows.

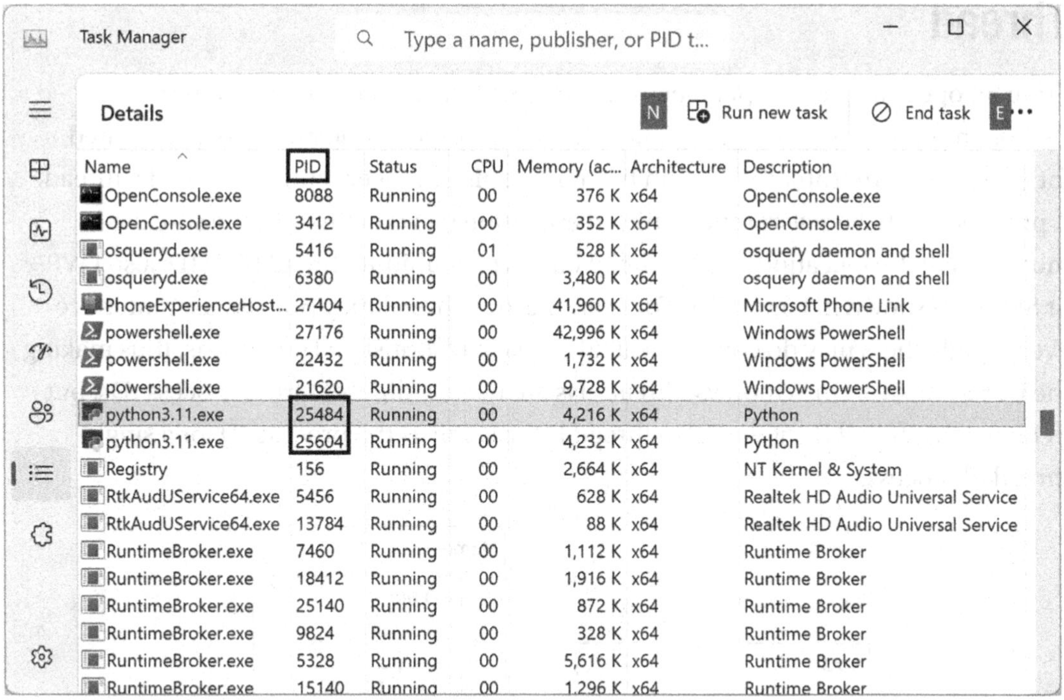

**Figure 1-2.** *Two running python3.11.exe processes on Windows*

On Linux, you can also see two processes when you execute the same script in two separate terminals:

```
~/Chapter1$ which python3
/usr/bin/python3
```

```
~/Chapter1$ ps -a
  PID TTY          TIME CMD
   17 pts/0    00:00:00 mc
   60 pts/2    00:00:00 python3
   61 pts/1    00:00:00 python3
   80 pts/3    00:00:00 ps
```

**Note**    The operating system controls hardware and processes/threads. From a high level, it is just a collection of processes with the operating system kernel as a process too.

# Thread

From an operating system perspective, a process is just a memory container for a
Python interpreter, its code, and data. But the interpreter code needs to be executed,
for example, to interpret the Python bytecode. This unit of execution is called a thread.
A process may have several such units of execution (several threads, the so-called
multithreaded application). Each thread has its own unique thread ID (TID, also LWP
or SPID), as shown in Figure 1-3. For example, one thread may process user interface
events and others may do complex calculations in response to UI requests, thus making
the UI responsive. On Windows, thread IDs are usually different from process IDs, but
in Linux, the thread ID of the main thread is the same as the process ID for a single-
threaded process.

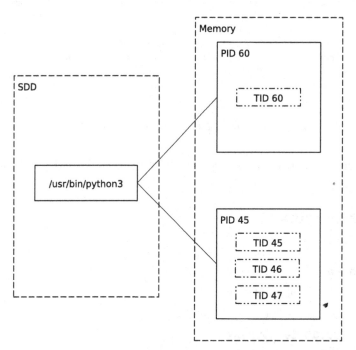

**Figure 1-3.** *Two python3 processes with different numbers of threads*

To model multithreading, I executed the code in Listing 1-2 on both Windows and Linux.

*Listing 1-2.*  A Simple Script to Model Multiple Threads

```python
import time
import threading

def thread_func():
    foo()

def main():
    t1 = threading.Thread(target=thread_func)
    t1.start()
    t2 = threading.Thread(target=thread_func)
    t2.start()
    t1.join()
    t2.join()

def foo():
    bar()

def bar():
    while True:
        time.sleep(1)

if __name__ == "__main__":
    main()
```

Figure 1-4 shows that in Windows, you can see 11 threads at the beginning (this number later changes to 7 and then to 5). You see that the number of threads may be greater than expected.

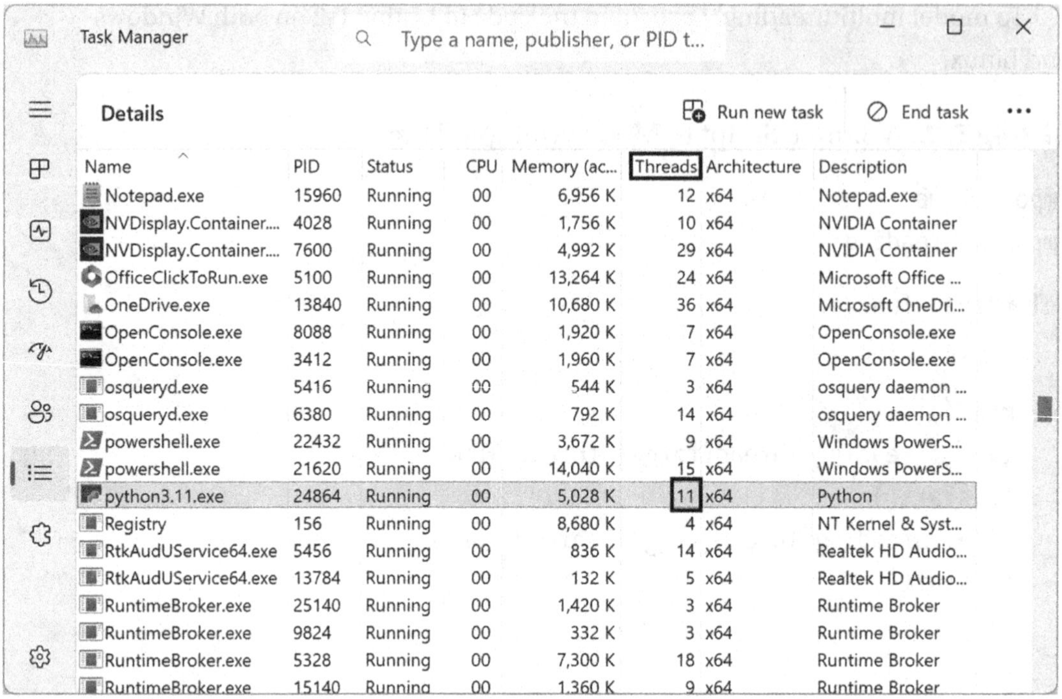

*Figure 1-4.*  *The number of threads in the running python3.11.exe process on Windows*

In Linux, you can see the expected number of threads – 3:

```
~/Chapter1$ ps -aT
  PID  SPID TTY           TIME CMD
   17    17 pts/0     00:00:00 mc
   45    45 pts/2     00:00:00 python3
   45    46 pts/2     00:00:00 python3
   45    47 pts/2     00:00:00 python3
   54    54 pts/1     00:00:00 ps
```

# Stack Trace (Backtrace, Traceback)

I should distinguish Python source code tracebacks (which we call *managed stack traces*) and unmanaged (native) ones from the Python compiler and interpreter that compiles to and executes Python byte code. You will see this distinction in some chapters for several case studies and how to get both traces. But, for now, I will just show the difference.

Listing 1-3 shows managed stack trace. Listing 1-4 shows the corresponding unmanaged Linux stack trace with debugging symbols (the most recent call first). Listing 1-5 shows the corresponding unmanaged Windows stack trace without debugging symbols (the most recent call first).

***Listing 1-3.*** Managed Stack Trace from the Execution of the Python Script from Listing 1-1

```
Traceback (most recent call last):
  File "process.py", line 14, in <module>
    main()
  File "process.py", line 4, in main
    foo()
  File "process.py", line 7, in foo
    bar()
  File "process.py", line 11, in bar
    time.sleep(1)
```

***Listing 1-4.*** Unmanaged Linux Backtrace from the Execution of the Python Script from Listing 1-1 with Debugging Symbols

```
#0  0x00007f6bc84e6b97 in __GI___select (nfds=0, readfds=0x0, writefds=0x0,
exceptfds=0x0, timeout=0x7ffc60288fe0)
    at ../sysdeps/unix/sysv/linux/select.c:41
#1  0x00000000004e8965 in pysleep (secs=<optimized out>) at ../Modules/
timemodule.c:1829
#2  time_sleep (self=<optimized out>, obj=<optimized out>, self=<optimized
out>, obj=<optimized out>)
    at ../Modules/timemodule.c:371
#3  0x00000000005d8711 in _PyMethodDef_RawFastCallKeywords (method=0x82dbe0
<time_methods+288>,
    self=<module at remote 0x7f6bc800dc78>, args=0x7f6bc80c4550,
    nargs=<optimized out>, kwnames=<optimized out>)
    at ../Objects/call.c:644
#4  0x000000000054b330 in _PyCFunction_FastCallKeywords (kwnames=<optimized
out>, nargs=<optimized out>,
```

```
       args=0x7f6bc80c4550, func=<built-in method sleep of module object at
       remote 0x7f6bc800dc78>)
       at ../Objects/call.c:730
#5   call_function (pp_stack=0x7ffc60289150, oparg=<optimized out>,
kwnames=<optimized out>) at ../Python/ceval.c:4568
#6   0x00000000005524cd in _PyEval_EvalFrameDefault (f=<optimized out>,
throwflag=<optimized out>)
       at ../Python/ceval.c:3093
#7   0x00000000005d91fc in PyEval_EvalFrameEx (throwflag=0,
       f=Frame 0x7f6bc80c43d8, for file process.py, line 11, in bar ()) at ../
Python/ceval.c:547
#8   function_code_fastcall (globals=<optimized out>, nargs=<optimized out>,
args=<optimized out>, co=<optimized out>)
       at ../Objects/call.c:283
#9   _PyFunction_FastCallKeywords (func=<optimized out>, stack=<optimized
out>, nargs=<optimized out>,
       kwnames=<optimized out>) at ../Objects/call.c:408
#10 0x000000000054e5ac in call_function (kwnames=0x0, oparg=<optimized
out>, pp_stack=<synthetic pointer>)
       at ../Python/ceval.c:4616
#11 _PyEval_EvalFrameDefault (f=<optimized out>, throwflag=<optimized out>)
at ../Python/ceval.c:3124
#12 0x00000000005d91fc in PyEval_EvalFrameEx (throwflag=0,
       f=Frame 0x7f6bc80105e8, for file process.py, line 7, in foo ()) at ../
       Python/ceval.c:547
#13 function_code_fastcall (globals=<optimized out>, nargs=<optimized out>,
args=<optimized out>, co=<optimized out>)
       at ../Objects/call.c:283
#14 _PyFunction_FastCallKeywords (func=<optimized out>, stack=<optimized
out>, nargs=<optimized out>,
       kwnames=<optimized out>) at ../Objects/call.c:408
--Type <RET> for more, q to quit, c to continue without paging--
#15 0x000000000054e5ac in call_function (kwnames=0x0, oparg=<optimized
out>, pp_stack=<synthetic pointer>)
       at ../Python/ceval.c:4616
```

```
#16 _PyEval_EvalFrameDefault (f=<optimized out>, throwflag=<optimized out>)
at ../Python/ceval.c:3124
#17 0x00000000005d91fc in PyEval_EvalFrameEx (throwflag=0, f=Frame
0x205ade8, for file process.py, line 4, in main ())
    at ../Python/ceval.c:547
#18 function_code_fastcall (globals=<optimized out>, nargs=<optimized out>,
args=<optimized out>, co=<optimized out>)
    at ../Objects/call.c:283
#19 _PyFunction_FastCallKeywords (func=<optimized out>, stack=<optimized
out>, nargs=<optimized out>,
    kwnames=<optimized out>) at ../Objects/call.c:408
#20 0x000000000054e5ac in call_function (kwnames=0x0, oparg=<optimized
out>, pp_stack=<synthetic pointer>)
    at ../Python/ceval.c:4616
#21 _PyEval_EvalFrameDefault (f=<optimized out>, throwflag=<optimized out>)
at ../Python/ceval.c:3124
#22 0x000000000054bcc2 in PyEval_EvalFrameEx (throwflag=0,
    f=Frame 0x7f6bc80ab9f8, for file process.py, line 14, in <module> ())
    at ../Python/ceval.c:547
#23 _PyEval_EvalCodeWithName (_co=<optimized out>, globals=<optimized out>,
locals=<optimized out>,
    args=<optimized out>, argcount=<optimized out>, kwnames=0x0,
    kwargs=0x0, kwcount=<optimized out>, kwstep=2,
    defs=0x0, defcount=0, kwdefs=0x0, closure=0x0, name=0x0, qualname=0x0)
    at ../Python/ceval.c:3930
#24 0x000000000054e0a3 in PyEval_EvalCodeEx (closure=0x0, kwdefs=0x0,
defcount=0, defs=0x0, kwcount=0, kws=0x0,
    argcount=0, args=0x0, locals=<optimized out>, globals=<optimized out>,
    _co=<optimized out>)
    at ../Python/ceval.c:3959
#25 PyEval_EvalCode (co=<optimized out>, globals=<optimized out>,
locals=<optimized out>) at ../Python/ceval.c:524
#26 0x0000000000630ce2 in run_mod (mod=<optimized out>,
filename=<optimized out>,
```

```
    globals={'__name__': '__main__', '__doc__': None, '__package__': None,
    '__loader__': <SourceFileLoader(name='__main__', path='process.py') at
    remote 0x7f6bc803dfd0>, '__spec__': None, '__annotations__': {}, '__
    builtins__': <module at remote 0x7f6bc8102c28>, '__file__': 'process.
    py', '__cached__': None, 'time': <module at remote 0x7f6bc800dc78>,
    'main': <function at remote 0x7f6bc80791e0>, 'foo': <function at remote
    0x7f6bc7f69c80>, 'bar': <function at remote 0x7f6bc7f69d08>},
    locals={'__name__': '__main__', '__doc__': None, '__package__': None,
    '__loader__': <SourceFileLoader(name='__main__', path='process.py')
    at remote 0x7f6bc803dfd0>, '__spec__': None, '__annotations__': {},
    '__builtins__': <module at rem--Type <RET> for more, q to quit, c to
    continue without paging--
ote 0x7f6bc8102c28>, '__file__': 'process.py', '__cached__': None,
'time': <module at remote 0x7f6bc800dc78>, 'main': <function at
remote 0x7f6bc80791e0>, 'foo': <function at remote 0x7f6bc7f69c80>,
'bar': <function at remote 0x7f6bc7f69d08>}, flags=<optimized out>,
arena=<optimized out>) at ../Python/pythonrun.c:1035
#27 0x0000000000630d97 in PyRun_FileExFlags (fp=0x2062390, filename_
str=<optimized out>, start=<optimized out>,
    globals={'__name__': '__main__', '__doc__': None, '__package__': None,
    '__loader__': <SourceFileLoader(name='__main__', path='process.py') at
    remote 0x7f6bc803dfd0>, '__spec__': None, '__annotations__': {}, '__
    builtins__': <module at remote 0x7f6bc8102c28>, '__file__': 'process.
    py', '__cached__': None, 'time': <module at remote 0x7f6bc800dc78>,
    'main': <function at remote 0x7f6bc80791e0>, 'foo': <function at remote
    0x7f6bc7f69c80>, 'bar': <function at remote 0x7f6bc7f69d08>},
    locals={'__name__': '__main__', '__doc__': None, '__package__': None,
    '__loader__': <SourceFileLoader(name='__main__', path='process.py') at
    remote 0x7f6bc803dfd0>, '__spec__': None, '__annotations__': {}, '__
    builtins__': <module at remote 0x7f6bc8102c28>, '__file__': 'process.
    py', '__cached__': None, 'time': <module at remote 0x7f6bc800dc78>,
    'main': <function at remote 0x7f6bc80791e0>, 'foo': <function at
    remote 0x7f6bc7f69c80>, 'bar': <function at remote 0x7f6bc7f69d08>},
    closeit=1, flags=0x7ffc6028989c) at ../Python/pythonrun.c:988
```

```
#28 0x00000000006319ff in PyRun_SimpleFileExFlags (fp=0x2062390,
filename=<optimized out>, closeit=1,
    flags=0x7ffc6028989c) at ../Python/pythonrun.c:429
#29 0x000000000065432e in pymain_run_file (p_cf=0x7ffc6028989c,
filename=<optimized out>, fp=0x2062390)
    at ../Modules/main.c:427
#30 pymain_run_filename (cf=0x7ffc6028989c, pymain=0x7ffc60289970) at ../
Modules/main.c:1627
#31 pymain_run_python (pymain=0x7ffc60289970) at ../Modules/main.c:2877
#32 pymain_main (pymain=<optimized out>, pymain=<optimized out>) at ../
Modules/main.c:3038
#33 0x000000000065468e in _Py_UnixMain (argc=<optimized out>,
argv=<optimized out>) at ../Modules/main.c:3073
#34 0x00007f6bc841a09b in __libc_start_main (main=0x4bc560 <main>, argc=2,
argv=0x7ffc60289ab8, init=<optimized out>,
    fini=<optimized out>, rtld_fini=<optimized out>, stack_
end=0x7ffc60289aa8) at ../csu/libc-start.c:308
#35 0x00000000005e0e8a in _start () at ../Modules/main.c:797
```

**Listing 1-5.** *Unmanaged Windows Stack Trace from the Execution of the Python Script from Listing 1-1 Without Debugging Symbols*

```
00 00000090`7e1ef0a8 00007ff9`8c44fcf9    ntdll!NtWaitForMultipleObjects+
0x14
01 00000090`7e1ef0b0 00007ff9`8c44fbfe    KERNELBASE!WaitForMultipleObject
sEx+0xe9
02 00000090`7e1ef390 00007ff8`ef943986    KERNELBASE!WaitForMultipleObject
s+0xe
03 00000090`7e1ef3d0 00007ff8`ef94383d    python311!PyTraceBack_Print_
Indented+0x35a
04 00000090`7e1ef430 00007ff8`ef81a6b2    python311!PyTraceBack_Print_
Indented+0x211
05 00000090`7e1ef460 00007ff8`ef82fa77    python311!PyEval_
EvalFrameDefault+0x8f2
06 00000090`7e1ef670 00007ff8`ef82f137    python311!PyMapping_Check+0x1eb
07 00000090`7e1ef6b0 00007ff8`ef82d80a    python311!PyEval_EvalCode+0x97
```

```
08 00000090`7e1ef730 00007ff8`ef82d786     python311!PyMapping_Items+0x11e
09 00000090`7e1ef760 00007ff8`ef97a17e     python311!PyMapping_Items+0x9a
0a 00000090`7e1ef7a0 00007ff8`ef7e33a5     python311!PyThread_tss_is_
created+0x53ce
0b 00000090`7e1ef810 00007ff8`ef8da620     python311!PyRun_
SimpleFileObject+0x11d
0c 00000090`7e1ef880 00007ff8`ef8daaef     python311!PyRun_
AnyFileObject+0x54
0d 00000090`7e1ef8b0 00007ff8`ef8dab5f     python311!Py_
MakePendingCalls+0x38f
0e 00000090`7e1ef980 00007ff8`ef8db964     python311!Py_
MakePendingCalls+0x3ff
0f 00000090`7e1ef9b0 00007ff8`ef8db7f5     python311!Py_RunMain+0x184
10 00000090`7e1efa20 00007ff8`ef8260d9     python311!Py_RunMain+0x15
11 00000090`7e1efa50 00007ff6`aefe1230     python311!Py_Main+0x25
12 00000090`7e1efaa0 00007ff9`8e1c26ad     python+0x1230
13 00000090`7e1efae0 00007ff9`8ef6a9f8     KERNEL32!BaseThreadInitThunk+0x
1d
14 00000090`7e1efb10 00000000`00000000     ntdll!RtlUserThreadStart+0x28
```

---

**Note**    Each thread has its own stack trace (backtrace).

---

# Symbol Files

Symbol files allow a debugger to map memory addresses to symbolic information such
as function and variable names. For example, if you download and apply symbol files to
the Windows example above, you get a much better and more accurate stack trace, as
shown in Listing 1-6.

**Listing 1-6.** Unmanaged Windows Stack Trace from the Execution of the Python Script from Listing 1-1 with Debugging Symbols

```
00 00000090`7e1ef0a8 00007ff9`8c44fcf9       ntdll!NtWaitForMultipleObjects+
0x14
01 00000090`7e1ef0b0 00007ff9`8c44fbfe       KERNELBASE!WaitForMultipleObject
sEx+0xe9
02 00000090`7e1ef390 00007ff8`ef943986       KERNELBASE!WaitForMultipleObject
s+0xe
03 00000090`7e1ef3d0 00007ff8`ef94383d       python311!pysleep+0x11a
04 00000090`7e1ef430 00007ff8`ef81a6b2       python311!time_sleep+0x2d
05 00000090`7e1ef460 00007ff8`ef82fa77       python311!_PyEval_
EvalFrameDefault+0x8f2
06 (Inline Function) --------`--------       python311!_PyEval_EvalFrame+0x1e
07 00000090`7e1ef670 00007ff8`ef82f137       python311!_PyEval_Vector+0x77
08 00000090`7e1ef6b0 00007ff8`ef82d80a       python311!PyEval_EvalCode+0x97
09 00000090`7e1ef730 00007ff8`ef82d786       python311!run_eval_code_obj+0x52
0a 00000090`7e1ef760 00007ff8`ef97a17e       python311!run_mod+0x72
0b 00000090`7e1ef7a0 00007ff8`ef7e33a5       python311!pyrun_file+0x196b66
0c 00000090`7e1ef810 00007ff8`ef8da620       python311!_PyRun_
SimpleFileObject+0x11d
0d 00000090`7e1ef880 00007ff8`ef8daaef       python311!_PyRun_
AnyFileObject+0x54
0e 00000090`7e1ef8b0 00007ff8`ef8dab5f       python311!pymain_run_file_
obj+0x10b
0f 00000090`7e1ef980 00007ff8`ef8db964       python311!pymain_run_file+0x63
10 00000090`7e1ef9b0 00007ff8`ef8db7f5       python311!pymain_run_
python+0x140
11 00000090`7e1efa20 00007ff8`ef8260d9       python311!Py_RunMain+0x15
12 00000090`7e1efa50 00007ff6`aefe1230       python311!Py_Main+0x25
13 (Inline Function) --------`--------       python!invoke_main+0x22
14 00000090`7e1efaa0 00007ff9`8e1c26ad       python!__scrt_common_main_
seh+0x10c
15 00000090`7e1efae0
00007ff9`8ef6a9f8       KERNEL32!BaseThreadInitThunk+0x1d
16 00000090`7e1efb10 00000000`00000000       ntdll!RtlUserThreadStart+0x28
```

# Module

Like the distinction between managed and unmanaged stack traces, there is a difference between Python modules (which may correspond to files in traceback) and native modules such as DLLs in Windows and .so files in Linux, which are loaded into memory when you execute the Python compiler/interpreter. For example, the following shared libraries are loaded in Linux for the simple multithreaded example from Listing 1-2:

```
~/Chapter1$ pmap 60
60:    python3 process.py
0000000000400000     132K r---- python3.7
0000000000421000    2256K r-x-- python3.7
0000000000655000    1712K r---- python3.7
0000000000801000       4K r---- python3.7
0000000000802000     664K rw--- python3.7
00000000008a8000     140K rw---   [ anon ]
0000000001fff000     660K rw---   [ anon ]
00007f6bc7f69000    1684K rw---   [ anon ]
00007f6bc810e000    2964K r---- locale-archive
00007f6bc83f3000      12K rw---   [ anon ]
00007f6bc83f6000     136K r---- libc-2.28.so
00007f6bc8418000    1308K r-x-- libc-2.28.so
00007f6bc855f000     304K r---- libc-2.28.so
00007f6bc85ab000       4K ----- libc-2.28.so
00007f6bc85ac000      16K r---- libc-2.28.so
00007f6bc85b0000       8K rw--- libc-2.28.so
00007f6bc85b2000      16K rw---   [ anon ]
00007f6bc85b6000      52K r---- libm-2.28.so
00007f6bc85c3000     636K r-x-- libm-2.28.so
00007f6bc8662000     852K r---- libm-2.28.so
00007f6bc8737000       4K r---- libm-2.28.so
00007f6bc8738000       4K rw--- libm-2.28.so
00007f6bc8739000       8K rw---   [ anon ]
00007f6bc873b000      12K r---- libz.so.1.2.11
00007f6bc873e000      72K r-x-- libz.so.1.2.11
00007f6bc8750000      24K r---- libz.so.1.2.11
```

```
00007f6bc8756000      4K -----  libz.so.1.2.11
00007f6bc8757000      4K r----  libz.so.1.2.11
00007f6bc8758000      4K rw---  libz.so.1.2.11
00007f6bc8759000     16K r----  libexpat.so.1.6.8
00007f6bc875d000    132K r-x--  libexpat.so.1.6.8
00007f6bc877e000     80K r----  libexpat.so.1.6.8
00007f6bc8792000      4K -----  libexpat.so.1.6.8
00007f6bc8793000      8K r----  libexpat.so.1.6.8
00007f6bc8795000      4K rw---  libexpat.so.1.6.8
00007f6bc8796000      4K r----  libutil-2.28.so
00007f6bc8797000      4K r-x--  libutil-2.28.so
00007f6bc8798000      4K r----  libutil-2.28.so
00007f6bc8799000      4K r----  libutil-2.28.so
00007f6bc879a000      4K rw---  libutil-2.28.so
00007f6bc879b000      4K r----  libdl-2.28.so
00007f6bc879c000      4K r-x--  libdl-2.28.so
00007f6bc879d000      4K r----  libdl-2.28.so
00007f6bc879e000      4K r----  libdl-2.28.so
00007f6bc879f000      4K rw---  libdl-2.28.so
00007f6bc87a0000     24K r----  libpthread-2.28.so
00007f6bc87a6000     60K r-x--  libpthread-2.28.so
00007f6bc87b5000     24K r----  libpthread-2.28.so
00007f6bc87bb000      4K r----  libpthread-2.28.so
00007f6bc87bc000      4K rw---  libpthread-2.28.so
00007f6bc87bd000     16K rw---   [ anon ]
00007f6bc87c1000      4K r----  libcrypt-2.28.so
00007f6bc87c2000     24K r-x--  libcrypt-2.28.so
00007f6bc87c8000      8K r----  libcrypt-2.28.so
00007f6bc87ca000      4K -----  libcrypt-2.28.so
00007f6bc87cb000      4K r----  libcrypt-2.28.so
00007f6bc87cc000      4K rw---  libcrypt-2.28.so
00007f6bc87cd000    192K rw---   [ anon ]
00007f6bc8801000     28K r--s-  gconv-modules.cache
00007f6bc8808000      4K r----  ld-2.28.so
00007f6bc8809000    120K r-x--  ld-2.28.so
```

```
00007f6bc8827000        32K  r----  ld-2.28.so
00007f6bc882f000         4K  r----  ld-2.28.so
00007f6bc8830000         4K  rw---  ld-2.28.so
00007f6bc8831000         4K  rw---  [ anon ]
00007ffc6026a000       132K  rw---  [ stack ]
00007ffc60356000        16K  r----  [ anon ]
00007ffc6035a000         4K  r-x--  [ anon ]
 total               14700K
```

The Windows version has the following loaded modules:

```
00007ff6`aefe0000 00007ff6`aeffa000   python    python.exe
00007ff8`ef7e0000 00007ff8`efdad000   python311 python311.dll
00007ff9`62950000 00007ff9`6296b000   VCRUNTIME140 VCRUNTIME140.dll
00007ff9`7f1e0000 00007ff9`7f1ea000   VERSION   VERSION.dll
00007ff9`8bce0000 00007ff9`8bd08000   bcrypt    bcrypt.dll
00007ff9`8c3f0000 00007ff9`8c793000   KERNELBASE KERNELBASE.dll
00007ff9`8c840000 00007ff9`8c951000   ucrtbase ucrtbase.dll
00007ff9`8c960000 00007ff9`8c9db000   bcryptprimitives bcryptprimitives.dll
00007ff9`8d150000 00007ff9`8d1c1000   WS2_32    WS2_32.dll
00007ff9`8d1d0000 00007ff9`8d2e7000   RPCRT4    RPCRT4.dll
00007ff9`8dd50000 00007ff9`8ddf7000   msvcrt    msvcrt.dll
00007ff9`8ded0000 00007ff9`8df7e000   ADVAPI32  ADVAPI32.dll
00007ff9`8e1b0000 00007ff9`8e272000   KERNEL32  KERNEL32.DLL
00007ff9`8e280000 00007ff9`8e324000   sechost   sechost.dll
00007ff9`8ef10000 00007ff9`8f124000   ntdll     ntdll.dll
```

# Memory Dump

A process memory can be saved in a memory dump file.

*An undigested and voluminous mass of information about a problem or the state of a system and most especially one consisting of hex runes describing the byte-by-byte state of memory.*

Eric S. Raymond, *The New Hacker's Dictionary*, Third Edition

These memory dumps are also called *core dumps* in Linux. It is also possible to get a kernel memory dump and a dump of physical memory (also called a *complete memory dump* in Windows). Figure 1-5 shows different memory dump types.

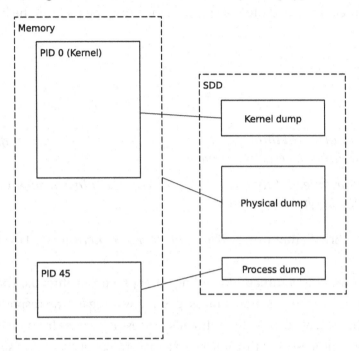

**Figure 1-5.** *Memory dump types*

Memory dumps may be useful for debugging hard-to-reproduce intermittent problems. This approach is called postmortem debugging. You will see some case studies in the following chapters.

# Crash

*To fail suddenly. "Has the system just crashed?" "Something crashed the OS!" Also used transitively to indicate the cause of the crash (usually a person or a program, or both). "Those idiots playing SPACEWAR crashed the system."*

Eric S. Raymond, *The New Hacker's Dictionary*, Third Edition.

When something illegal happens inside a process thread, such as when memory outside its available range is accessed or you write to read-only memory, the operating system reports the error and terminates the process. It may also save the process memory into a memory dump file. The process then disappears from the list of available processes.

# Hang

*1. To wait for an event that will never occur. "The system is hanging because it can't read from the crashed drive."*

*2. To wait for an event to occur. "The program displays a menu and then hangs until you type a character."*

Eric S. Raymond, *The New Hacker's Dictionary*, Third Edition

Threads interact with other threads, including other processes' threads. These interactions can be viewed as sending messages and waiting for the responses. Some processes may be critical because their threads process messages from many other threads from other processes. If threads from such a critical process stop sending responses, all other waiting threads are blocked. A deadlock is when two threads are waiting for each other. When hanging, the process continues to be present in the list of available processes. There are also processes (critical components) that, when their threads hang, block threads from many other processes (noncritical components). Figure 1-6 depicts such components and their interaction abstracted via messages in the normal scenario, and Figure 1-7 shows the abnormal scenario when noncritical components are blocked and waiting for responses because the critical components are deadlocked.

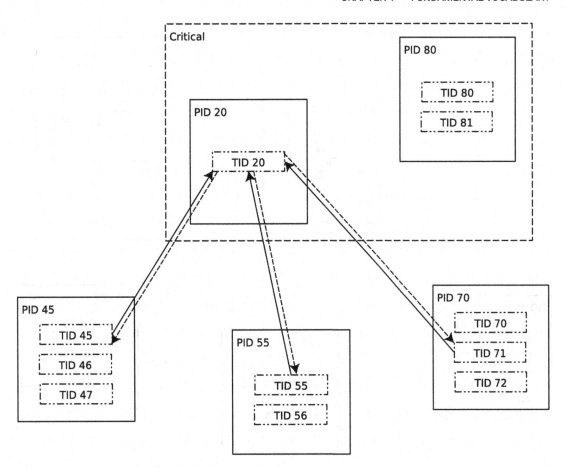

***Figure 1-6.*** *Request and response interaction between critical and noncritical system components*

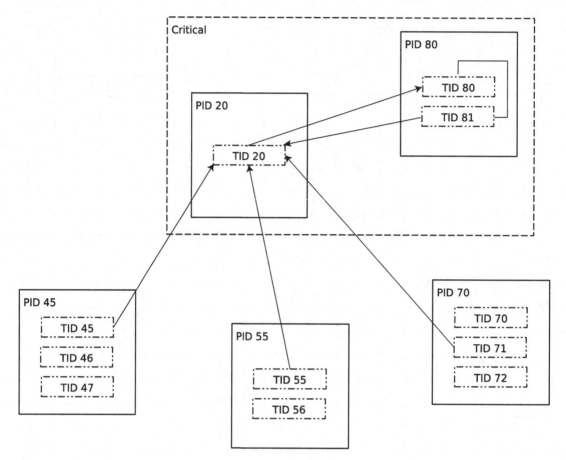

**Figure 1-7.** *Blocked and deadlocked components*

## Summary

In this chapter, you learned the fundamental vocabulary you will use in subsequent chapters. The next chapter introduces a pattern-oriented debugging approach.

# CHAPTER 2

# Pattern-Oriented Debugging

This chapter introduces the pattern-oriented debugging process approach and the pattern languages you will use in subsequent chapters.

## The History of the Idea

The idea of using patterns in debugging is not new[1]. Earlier, such patterns came in two types: bug patterns[2] and debug patterns[3]. Before 2000, only a few debugging-related patterns could be found, such as the Debug Printing Method[4].

Bug patterns are usually specific patterns for specific languages and platforms. By bugs, we mean software defects. Usually, these are related to the source code but can also be related to configuration and data models.

Using source code as a starting point for debugging is only possible for a limited number of scenarios, such as when you have a Python stack trace. However, there are many cases when the starting point for source code investigation is unknown. Here, a well-defined process may benefit. A number of debugging processes were proposed in the past, including multidisciplinary approaches[5].

---

[1] Mehdi Amoui et al., "A Pattern Language for Software Debugging," International Journal of Computer Science, vol. 1, no. 3, pp. 218-224, 2006. https://stargroup.uwaterloo.ca/~mamouika/papers/pdf/IJCS.2006.pdf

[2] Eric Allen, *Bug Patterns in Java*, 2002 (ISBN-13: 978-1590590614)

[3] https://en.wikipedia.org/wiki/Debugging_pattern

[4] Linda Rising, *The Pattern Almanac 2000*, p. 154 (ISBN-13: 978-0201615678)

[5] Robert Charles Metzger, *Debugging by Thinking: A Multidisciplinary Approach*, 2003 (ISBN-13: 978-1555583071)

The phrase "pattern-oriented debugging" appeared around 1987-1988 in the context of patterns of process interaction[6]. It is not the same as the "pattern-oriented debugging process" proposed in 2014[7] as further development of *unified debugging patterns* that were introduced in 2010[8]. Since then, these patterns have been used for successfully teaching Windows debugging of unmanaged (native, Win64, C, C++) and managed (.NET, C#) code[9] for almost a decade, starting in 2013. Overall, this pattern-oriented approach can be traced to our earlier presentation published as a book in 2011[10]. In it, we apply the same pattern-oriented process to Python debugging in cloud and machine learning environments.

# Patterns and Analysis Patterns

Before looking at the debugging process, a few words about patterns in the context of diagnostics and debugging. By a *pattern*, we mean a common recurrent identifiable set of indicators (symptoms, signs). By an *analysis* pattern, we mean a common recurrent analysis technique and method of pattern identification in a specific context. By *pattern language*, we mean common names of patterns and analysis patterns used for communication.

# Development Process

Let's first look at the traditional software development process stages. Figure 2-1 abstracts them from several development processes, including waterfall and iterative ones.

---

[6] Alfred A. Hough and Janice E. Cuny, "Initial experiences with a pattern-oriented parallel debugger." PADD '88: Proceedings of the 1988 ACM SIGPLAN and SIGOPS workshop on Parallel and distributed debugging November 1988 Pages 195–205 https://doi.org/10.1145/68210.69234

[7] Dmitry Vostokov, *Pattern-Oriented Debugging Process, in Theoretical Software Diagnostics: Collected Articles,* Third Edition, 2020 (ISBN-13: 978-1912636334), pp. 197-199

[8] Ibid., "Analysis, Architectural, Design, Implementation and Usage Debugging Patterns," p. 129

[9] *Accelerated Windows Debugging 4D*, Third Edition, 2022 (ISBN-13: 978-1912636532)

[10] Dmitry Vostokov, *Introduction to Pattern-Driven Software Problem Solving*, 2011 (ISBN-13: 978-1908043177)

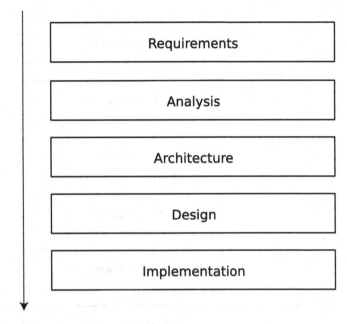

**Figure 2-1.** *Stages of the typical software development process*

# Development Patterns

For each stage, there exists some pattern language such as a vocabulary of solutions to common recurrent identifiable problems with grammar, semantics, and pragmatics. Figure 2-2 also includes software usage and presentation patterns for human-computer interaction. In this book, I assume you have familiarity with such pattern languages (some references are provided below).

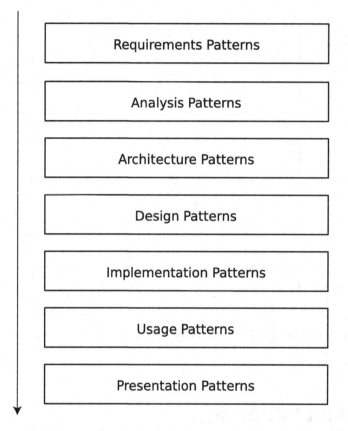

*Figure 2-2.* *Typical software development pattern languages*

# Debugging Process and Patterns

The debugging process mirrors the development process and development patterns, as shown in Figure 2-3. Let's look at each stage.

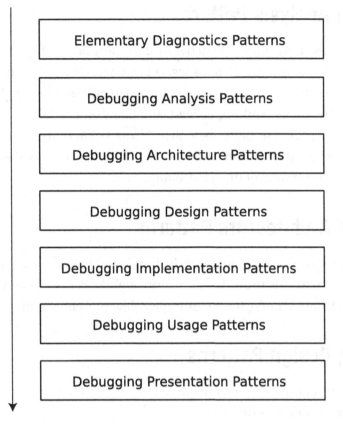

**Figure 2-3.** *Stages of the pattern-oriented debugging process*

# Elementary Diagnostics Patterns

Elementary software diagnostics patterns got inspiration from the *Elemental Design Patterns* book title[11], but they are different and correspond to requirements from the software development process. These are what software users experience, and the requirement is to eliminate such experiences via debugging.

---

[11] Jason McC. Smith, *Elemental Design Patterns*, 2012 (ISBN-13: 978-0321711922)

# Debugging Analysis Patterns

You need to diagnose the right problem before doing any debugging. Debugging analysis patterns correspond to software diagnostics. For example, in memory dump analysis, there are analysis patterns such as **Managed Code Exception**, **Managed Stack Trace**, **Stack Overflow**, **Deadlock**, **Spiking Thread**, and many others. There are hundreds of them[12]. Trace and log analysis patterns such as **Thread of Activity**, **Discontinuity**, **Time Delta**, **Counter Value**, **State Dump**, and many more are also included in this category[13]. We look at the most common of them in Chapter 4.

# Debugging Architecture Patterns

Debugging architecture patterns are partly inspired by POSA[14], for example, **Debug Event Subscription/Notification**. They are more high-level than design patterns that may differ for specific technologies, for example, object-oriented and functional.

# Debugging Design Patterns

Debugging design patterns are partly inspired by the GoF design pattern approach[15], for example, **Punctuated Execution**.

Both debugging architecture and debugging design patterns pertain to the development of debugging tools and to actual debugging architectures and designs as reusable solutions to common recurrent debugging problems in specific contexts.

---

[12] Dmitry Vostokov, *Encyclopedia of Crash Dump Analysis Patterns: Detecting Abnormal Software Structure and Behavior in Computer Memory*, Third Edition, 2020 (ISBN-13: 978-1912636303)

[13] Dmitry Vostokov, *Fundamentals of Trace and Log Analysis: A Pattern-Oriented Approach to Monitoring, Diagnostics, and Debugging*, Apress, 2023 (ISBN-13: 978-1484298954) and Dmitry Vostokov, *Trace, Log, Text, Narrative, Data: An Analysis Pattern Reference for Information Mining, Diagnostics, Anomaly Detection*, Fifth Edition, 2023 (ISBN-13: 978-1912636587)

[14] Frank Buschmann et al., *Pattern-Oriented Software Architecture: A System of Patterns*, 1996 (ISBN-13: 978-0471958697)

[15] Erich Gamma et al., *Design Patterns: Elements of Reusable Object-Oriented Software*, 1995 (ISBN-13: 978-0201633610)

## Debugging Implementation Patterns

Debugging implementation patterns are patterns of debugging strategies and core debugging techniques, such as **Break-in**, **Code Breakpoint**, **Data Breakpoint**, and others covered in subsequent chapters in Python debugging case studies.

## Debugging Usage Patterns

Debugging usage patterns are about debugging pragmatics of reusable debugging scenarios: how, what, when, and where to use the previous debugging pattern categories, such as using data breakpoints in user (process) and kernel space debugging.

## Debugging Presentation Patterns

Debugging presentation patterns are about user interface and interaction design[16], for example, watch dialog. These patterns are also related to debugging usage. We cover such patterns in the chapter devoted to existing Python IDEs and their usage for Python debugging.

In our opinion, recent Python debugging books[17] correspond to debugging implementation, usage, and presentation. Automated debugging[18] belongs to debugging architecture and design. In this book, we extract some of this knowledge into corresponding debugging pattern languages, combine them with pattern-oriented software diagnostics, and form a novel pattern-oriented Python debugging approach in the context of machine learning and cloud computing environments.

---

[16] Jan Borchers, *A Pattern Approach to Interaction Design*, 2001 (ISBN-13: 978-0471498285)

[17] Kristian Rother, *Pro Python Best Practices: Debugging, Testing and Maintenance*, 2017 (ISBN-13: 978-1484222409) and R. L. Zimmerman, *Python Debugging Handbook*, 2020 (ISBN-13: 979-8610761725)

[18] Andreas Zeller, *The Debugging Book: Tools and Techniques for Automated Software Debugging*, www.debuggingbook.org/ (*uses Python for examples*)

# Summary

In this chapter, you learned about the pattern-oriented debugging process, its stages, and corresponding pattern languages. The subsequent chapters provide examples of debugging patterns from each category from a Python programming perspective. The next chapter is about the first stage and its patterns: elementary diagnostics patterns.

# CHAPTER 3

# Elementary Diagnostics Patterns

In the previous chapter, I introduced the pattern-oriented debugging process with elementary (software) diagnostics patterns of abnormal software behavior that affect users and trigger software diagnostics and debugging if necessary (Figure 3-1).

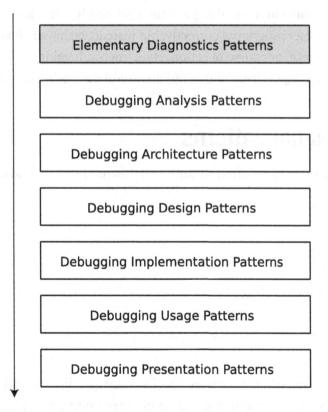

**Figure 3-1.** *Pattern-oriented debugging process and elementary diagnostics patterns*

© Dmitry Vostokov 2024
D. Vostokov, *Python Debugging for AI, Machine Learning, and Cloud Computing,*
https://doi.org/10.1007/978-1-4842-9745-2_3

There are only a few such patterns divided into two groups. The goal is to have the absolute minimum of them. In this chapter, you will look at them in more detail, learn how to recognize them during software execution, and collect relevant software execution artifacts. There are two groups of such patterns: functional and non-functional.

# Functional Patterns

The first group (functional) contains only one elementary diagnostics pattern: **Use-case Deviation**.

## Use-Case Deviation

By *a use case*, we mean a functional requirement, something that software is supposed to do for users (or other programs) correctly, but instead we have a deviation from the correct result, response, or course of action. Here tracing, logging, and running the program under a debugger (live debugging) are useful debugging techniques.

# Non-Functional Patterns

The second group (non-functional) contains a few elementary diagnostics patterns:

- Crash

- Hang

- Counter Value

- Error Message

# Crash

A **Crash** manifests itself when a software process suddenly disappears from the list of running processes. This list can be the output of the ps command in Linux, Activity Monitor in macOS, or the Task Manager detailed process list in Windows. The crash can happen from some Python exception, Python runtime, or an unhandled OS exception from some native library. For latter cases, it is recommended to configure the system to save *process memory dumps* (usually called *core dumps* in Linux and macOS and *crash dumps* in Windows). These memory dumps may help to see backtraces (stack traces) in case there is no output on the terminal (or console).

Saved memory dumps are loaded into a debugger to see the stack trace and program variables (memory values) at the time of the crash (postmortem debugging).

You can also run a program under a debugger from the start or attach a debugger before a crash. The debugger will also show a stack trace that led to a problem before the process disappeared (live debugging).

## How to Enable Process Core Dumps on Linux

Core dumps can be temporarily enabled for the current user using this command:

```
$ ulimit -c unlimited
```

Core dumps can be permanently enabled for every user except root by editing the /etc/security/limits.conf file. Add or uncomment the following line:

```
*       soft   core   unlimited
```

To limit the root to 1GB, add or uncomment the following line:

```
*       hard   core   1000000
```

## How to Enable Process Memory Dumps on Windows

This is best done via the LocalDumps registry key. The article *Collecting User-Mode Dumps*[1] at https://learn.microsoft.com/en-us/windows/win32/wer/collecting-user-mode-dumps outlines how to do it.

# Hang

A **Hang** manifests itself when a process becomes unresponsive (frozen) from a user's perspective (and also from the perspective of another process that may request some functionality). But it is still visible from the process list command or GUI process managers, which may or may not indicate that the process is frozen. This elementary pattern also includes delays. Here API and library tracing may point to the blocked call,

---

[1] https://learn.microsoft.com/en-us/windows/win32/wer/collecting-user-mode-dumps

or better, a manual memory dump should definitely show blocked process threads and even suggest whether there is a deadlock (threads mutually waiting for each other), the so-called postmortem debugging scenario.

It is also possible to start a program under a debugger (or attach it after the start), wait until it hangs, and then inspect threads and memory (live debugging).

## How to Generate Process Core Dumps on Linux

There are several methods:

- The `kill` command (requires `ulimit`)

  ```
  $ kill -s SIGQUIT PID
  $ kill -s SIGABRT PID
  ```

- gcore

  ```
  $ gcore [-o filename] PID
  ```

- ProcDump[2]

## How to Generate Process Memory Dumps on Windows

If you use a GUI environment, this can be done via Task Manager by selecting a Python process, right-clicking, and choosing Create dump file, as shown in Figure 3-2.

---

[2]https://github.com/Sysinternals/ProcDump-for-Linux

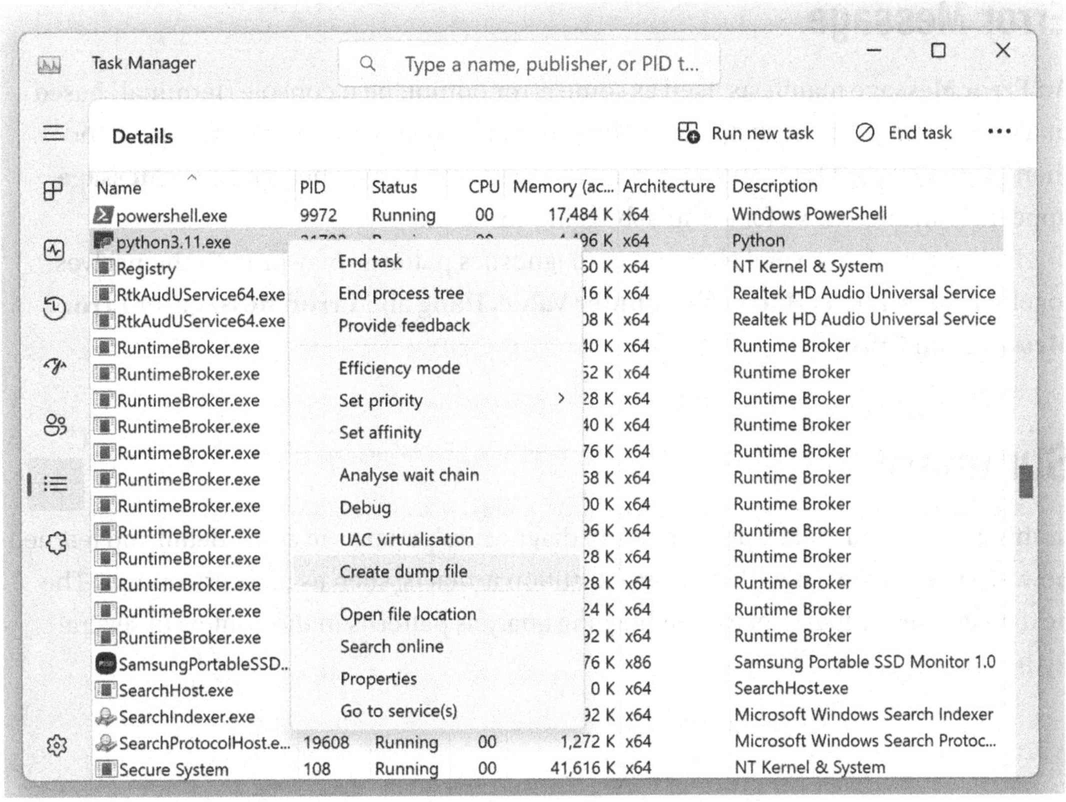

**Figure 3-2.** *Saving a process memory dump using Task Manager*

The ProcDump[3] command-line tool can be used for console environments. It is also recommended to use the -ma switch to save full process dumps.

# Counter Value

A **Counter Value** is about some measured quantity (metrics) or system variable that suddenly shows an unexpected (or abnormal) value. Therefore, it also includes resource leaks such as handles and memory plus CPU spikes. Here it is possible to instrument a program with additional code that logs data related to the **Counter Value**. It is also possible to run the program under a debugger and periodically stop it to inspect related variables and memory values (live debugging). Saving the memory dump periodically or after some threshold has the same effect (postmortem debugging).

---

[3] https://learn.microsoft.com/en-us/sysinternals/downloads/procdump

# Error Message

An **Error Message** manifests itself as some error output, be it console (terminal) based or a GUI message box or dialog box. The program may be run under a debugger and then get interrupted for inspection. Saving a memory dump when an **Error Message** appears is an alternative for some debugging scenarios.

Please note that several elementary diagnostics patterns may present themselves together, for example, **Hang** and **Counter Value, Hang** and **Error Message,** or **Error Message** and **Crash**.

# Summary

In this chapter, you looked at elementary diagnostics patterns in some detail and learned how to gather some types of software execution artifacts, such as memory dumps. The next chapter introduces specific debugging analysis patterns in the context of several Python case studies.

# CHAPTER 4

# Debugging Analysis Patterns

In the previous chapter, you looked at elementary diagnostics patterns that guide you in gathering, where appropriate, the necessary software execution artifacts, such as memory dumps, traces, and logs. You analyzed these artifacts for diagnostic indicators (signals, signs, and symptoms) of abnormal software structure and behavior that describe problems. We call common recurrent identifiable problems, together with a set of recommendations and possible solutions to apply in a specific context, as *diagnostic patterns*.

In this chapter, you will look at a pattern language of recurrent analysis techniques and methods of diagnostic pattern identification in a specific context: the domain of software diagnostics proper. This pattern language is independent of a particular debugger, language, or platform, such as **Stack Trace Collection**, but the debugger commands to gather and analyze the information may be different for Windows and Linux, and also for Python code stack trace and the stack trace of a Python process that executes Python code. We call such analysis pattern variants as *analysis pattern implementations*. There are many such analysis patterns[1] originally developed by the Software Diagnostics Institute[2], and they can be classified into groups. You will look at the most frequent ones.

---

[1] For native operating system diagnostics, there can be hundreds of them. See the book I referenced in Chapter 2: Dmitry Vostokov, *Encyclopedia of Crash Dump Analysis Patterns: Detecting Abnormal Software Structure and Behavior in Computer Memory*, Third Edition, 2020 (ISBN-13: 978-1912636303)

[2] https://dumpanalysis.org

© Dmitry Vostokov 2024
D. Vostokov, *Python Debugging for AI, Machine Learning, and Cloud Computing*,
https://doi.org/10.1007/978-1-4842-9745-2_4

Since software diagnostics precedes debugging, we call these diagnostic analysis patterns that trigger debugging architecture, design, and implementation as *debugging analysis patterns* (Figure 4-1).

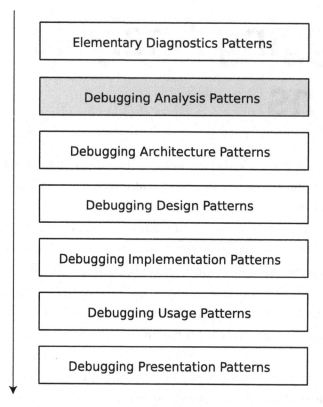

***Figure 4-1.*** *Pattern-oriented debugging process and debugging analysis patterns*

# Paratext

The name of this pattern was borrowed from literary theory[3]. In our context, it means additional supporting material for main software diagnostic artifacts, for example, operating system traces and logs showing how your Python process interacted with other processes and files plus its memory, handle, and CPU usage. In Windows, it can be the logs from Process Monitor[4]; in Linux, it's the output from the top, pmap, strace,

---

[3] https://en.wikipedia.org/wiki/Paratext

[4] https://learn.microsoft.com/en-us/sysinternals/downloads/procmon

ltrace, and lsof tools. For example, CPU thread usage information is not available in Linux process core memory dumps but is available in Windows process memory dumps, as you will see later for the **Spiking Thread** analysis pattern in the case study at the end of this chapter.

# State Dump

This pattern comes from trace and log analysis patterns[5] and is specifically useful for the scripting nature of Python code. You identify variables and objects of interest and periodically (or inside some function of interest) dump their values into a log file.

# Counter Value

This pattern also comes from trace and log analysis and is often related to metric monitoring, such as the size of memory consumption and the percentage of CPU used. These periodic values usually come from **Paratext** but may also be a part of a **State Dump**.

# Stack Trace Patterns

The first group of stack trace patterns contains the most important analysis patterns for debugging, which are patterns related to stack traces (backtraces, tracebacks):

- Stack Trace
- Runtime Thread
- Managed Stack Trace
- Stack Trace Collection
- Stack Trace Set

---

[5] See the book I referenced in Chapter 2: Dmitry Vostokov, *Trace, Log, Text, Narrative, Data: An Analysis Pattern Reference for Information Mining, Diagnostics, Anomaly Detection*, Fifth Edition, 2023 (ISBN-13: 978-1912636587)

# Stack Trace

In Python reference implementation (CPython), code is compiled into bytecode, and this bytecode is executed by the Python virtual machine. This virtual machine has its own functions that call other functions, including operating system APIs. The chain of such function calls at any given time is usually called a stack trace or backtrace.

The **Stack Trace** analysis pattern deals with methods to get such stack traces from the native debugger, for example, WinDbg for Windows and GDB for Linux, or from external stack trace sampling tools. This pattern is not limited to Python code execution thread; it covers any thread from a Python process and may include a stack trace from the kernel space.

# Runtime Thread

A Python process may be single-threaded or multithreaded. Windows processes usually have more threads than Linux ones. The **Runtime Thread** analysis pattern deals with the recognition of threads that are directly related to Python code execution, be it script code or garbage collection.

# Managed Stack Trace

The name **Managed** originally comes from Microsoft .NET managed code, which is compiled into an intermediate language that runs under language runtime. The same happens with Python source code. Regardless of whether it is interpreted or directly compiled, the chain of Python function calls constitutes its own stack trace.

The **Manage Stack Trace** analysis pattern deals with methods to get such stack traces (tracebacks). Usually, you capture them from console output or error logs when exceptions happen. However, scripts may have code added to periodically dump tracebacks in methods of interest or by using decorators. Listing 4-1 shows how to use the traceback module and decorators to dump a stack trace when some function of interest is used (see Chapter 6 for another example).

***Listing 4-1.***  A Script to Illustrate Traceback Module Usage

```
# managed-stack-trace.py

import traceback

def main():
    foo()

def foo():
    bar()

def managed_stack_trace(func):
    def call(*args, **kwargs):
        traceback.print_stack()
        return func(*args, **kwargs)
    return call

@managed_stack_trace
def bar():
    print("Hello Traceback!")

if __name__ == "__main__":
    main()
```

You get the following output when you call the bar function:

```
~/Python-Book/Chapter 4$ python3 managed-stack-trace.py
  File "managed-stack-trace.py", line 22, in <module>
    main()
  File "managed-stack-trace.py", line 6, in main
    foo()
  File "managed-stack-trace.py", line 9, in foo
    bar()
  File "managed-stack-trace.py", line 13, in call
    traceback.print_stack()
Hello Traceback!
```

If you don't want to print the decorator stack trace frame (highlighted in bold), you can use the inspect module to specify the starting frame for the traceback, as shown in Listing 4-2.

***Listing 4-2.*** A Script to Illustrate Inspecting Module Usage

```python
# managed-stack-trace2.py

import traceback
import inspect

def main():
    foo()

def foo():
    bar()

def managed_stack_trace(func):
    def call(*args, **kwargs):
        traceback.print_stack(f=inspect.currentframe().f_back)
        return func(*args, **kwargs)
    return call

@managed_stack_trace
def bar():
    print("Hello Traceback!")

if __name__ == "__main__":
    main()
```

You now get the improved output:

```
~/Python-Book/Chapter 4$ python3 managed-stack-trace2.py
  File "managed-stack-trace2.py", line 23, in <module>
    main()
  File "managed-stack-trace2.py", line 7, in main
    foo()
  File "managed-stack-trace2.py", line 10, in foo
    bar()
Hello Traceback!
```

Some native debuggers, such as GDB, have extensions to dump Python tracebacks from core memory dumps or when doing live native debugging. You will use one such extension in the case study at the end of this chapter.

## Source Stack Trace

Python **Managed Stack Traces** are identical to **Source Stack Traces** since they include file names and line numbers. However, when you have **Stack Traces** from Python runtime, they may only have source code information if you use a debug build of Python interpreter.

## Stack Trace Collection

The **Stack Trace Collection** analysis pattern deals with methods to get stack traces from all process threads or from several processes, or from the same process periodically, irrespective of whether these threads are **Runtime Threads** or not.

This pattern has different variants; for example, methods for getting all **Managed Stack Traces** are different from methods for getting all unmanaged **Stack Traces**. You will see this in the case study at the end of this chapter.

## Stack Trace Set

The **Stack Trace Set** analysis pattern deals with methods to get a subset having some property from **Stack Trace Collection**, for example, having some common function or just a set of unique non-duplicated stack traces.

# Exception Patterns

The second group of the most important analysis patterns for debugging is exception related:

- Managed Code Exception

- Nested Exception

- Exception Stack Trace

- Software Exception

# Managed Code Exception

The name **Managed** also comes from Microsoft .NET, like in the **Managed Stack Trace** analysis pattern. These are exceptions from Python code, and they have associated tracebacks.

# Nested Exception

The **Nested Exception** analysis pattern deals with cases when the code processing an exception raises a new exception. When the latter exception is caught, its data is also linked to the original exception, as shown in Listing 4-3.

***Listing 4-3.*** A Script to Illustrate Nested Exceptions

```python
# nested-exception.py

def main():
    foo()

def foo():
    bar()

def bar():
    try:
        raise Exception("Inner Exception")
    except Exception:
        raise Exception("Outer Exception")

if __name__ == "__main__":
    main()
```

When you run the script, you get the following output:

```
~/Python-Book/Chapter 4$ python3 nested-exception.py
Traceback (most recent call last):
  File "nested-exception.py", line 11, in bar
    raise Exception("Inner Exception")
Exception: Inner Exception
```

During the handling of the above exception, another exception occurred:

```
Traceback (most recent call last):
  File "nested-exception.py", line 16, in <module>
    main()
  File "nested-exception.py", line 4, in main
    foo()
  File "nested-exception.py", line 7, in foo
    bar()
  File "nested-exception.py", line 13, in bar
    raise Exception("Outer Exception")
Exception: Outer Exception
```

## Exception Stack Trace

When you have an exception, you want to get its **Exception Stack Trace** to analyze the sequence of function calls. A **Managed Code Exception** has a **Managed Stack Trace,** and depending on an artifact, it may be possible to extract an unmanaged runtime **Stack Trace**.

## Software Exception

The **Software Exception** analysis pattern is a general analysis pattern describing all sorts of non-hardware exceptions coming from either Python code or the Python runtime and from interfaces with non-Python modules and operating system API libraries.

## Module Patterns

When you talk about modules, you need to distinguish between Python modules and traditional operating system modules. A Python module is either a single file or a collection of files in a package. An operating system module is usually a shared library exporting an API and dynamically linked to an executable, for example .DLL files (Windows) or .so files (Linux).

# Module Collection

The **Module Collection** analysis pattern describes how to get all imported modules for a Python program at some point in time, such as at the time of the exception. Listing 4-4 shows how this can be done.

*Listing 4-4.* A Script to Illustrate Module Collection

```python
# module-collection.py

import math
import mymodule

def main():
    foo()

def foo():
    bar()

def bar():
    mymodule.myfunc()
    math.sqrt(-1)

if __name__ == "__main__":
    try:
        main()
    except:
        import sys
        for name in sys.modules:
            print(f"{sys.modules[name]}")

# mymodule.py
import random

def myfunc():
    random.seed()
```

If you run the script, you get the list of imported modules including modules imported by imported modules and their location:

```
~/Python-Book/Chapter 4$ python3 module-collection.py
<module 'sys' (built-in)>
<module 'builtins' (built-in)>
<module 'importlib._bootstrap' (frozen)>
<module '_imp' (built-in)>
<module '_warnings' (built-in)>
<module 'io' (built-in)>
<module 'marshal' (built-in)>
<module 'posix' (built-in)>
<module 'importlib._bootstrap_external' (frozen)>
<module '_thread' (built-in)>
<module '_weakref' (built-in)>
<module 'time' (built-in)>
<module 'zipimport' (frozen)>
<module '_codecs' (built-in)>
<module 'codecs' from '/usr/lib/python3.8/codecs.py'>
<module 'encodings.aliases' from '/usr/lib/python3.8/encodings/aliases.py'>
<module 'encodings' from '/usr/lib/python3.8/encodings/__init__.py'>
<module 'encodings.utf_8' from '/usr/lib/python3.8/encodings/utf_8.py'>
<module '_signal' (built-in)>
<module '__main__' from 'module-collection.py'>
<module 'encodings.latin_1' from '/usr/lib/python3.8/encodings/latin_1.py'>
<module '_abc' (built-in)>
<module 'abc' from '/usr/lib/python3.8/abc.py'>
<module 'io' from '/usr/lib/python3.8/io.py'>
<module '_stat' (built-in)>
<module 'stat' from '/usr/lib/python3.8/stat.py'>
<module '_collections_abc' from '/usr/lib/python3.8/_collections_abc.py'>
<module 'genericpath' from '/usr/lib/python3.8/genericpath.py'>
<module 'posixpath' from '/usr/lib/python3.8/posixpath.py'>
<module 'posixpath' from '/usr/lib/python3.8/posixpath.py'>
<module 'os' from '/usr/lib/python3.8/os.py'>
<module '_sitebuiltins' from '/usr/lib/python3.8/_sitebuiltins.py'>
```

```
<module '_locale' (built-in)>
<module '_bootlocale' from '/usr/lib/python3.8/_bootlocale.py'>
<module 'types' from '/usr/lib/python3.8/types.py'>
<module 'importlib._bootstrap' (frozen)>
<module 'importlib._bootstrap_external' (frozen)>
<module 'warnings' from '/usr/lib/python3.8/warnings.py'>
<module 'importlib' from '/usr/lib/python3.8/importlib/__init__.py'>
<module 'importlib.machinery' from '/usr/lib/python3.8/importlib/
machinery.py'>
<module 'importlib.abc' from '/usr/lib/python3.8/importlib/abc.py'>
<module '_operator' (built-in)>
<module 'operator' from '/usr/lib/python3.8/operator.py'>
<module 'keyword' from '/usr/lib/python3.8/keyword.py'>
<module '_heapq' (built-in)>
<module 'heapq' from '/usr/lib/python3.8/heapq.py'>
<module 'itertools' (built-in)>
<module 'reprlib' from '/usr/lib/python3.8/reprlib.py'>
<module '_collections' (built-in)>
<module 'collections' from '/usr/lib/python3.8/collections/__init__.py'>
<module '_functools' (built-in)>
<module 'functools' from '/usr/lib/python3.8/functools.py'>
<module 'contextlib' from '/usr/lib/python3.8/contextlib.py'>
<module 'importlib.util' from '/usr/lib/python3.8/importlib/util.py'>
<module 'zope' from '/usr/lib/python3/dist-packages/zope/__init__.py'>
<module 'apport_python_hook' from '/usr/lib/python3/dist-packages/apport_
python_hook.py'>
<module 'sitecustomize' from '/usr/lib/python3.8/sitecustomize.py'>
<module 'site' from '/usr/lib/python3.8/site.py'>
<module 'math' (built-in)>
<module '_bisect' (built-in)>
<module 'bisect' from '/usr/lib/python3.8/bisect.py'>
<module '_sha512' (built-in)>
<module '_random' (built-in)>
<module 'random' from '/usr/lib/python3.8/random.py'>
<module 'mymodule' from '/home/ubuntu/Python-Book/Chapter 4/mymodule.py'>
```

For an executable program, a module collection is usually a collection of all loaded DLLs or shared libraries. You will learn how to get this information when you analyze problems of interfacing with non-Python modules and an operating system using debuggers such as WinDbg and GDB.

## Not My Version

The name of this pattern originates from the NotMyFault[6] Sysinternals tool for modeling crashes, hangs, and leaks on Windows platforms. When looking at **Module Collection** information, you may spot module or package versions, not the ones you intended.

## Exception Module

A **Managed Code Exception** happens in a function from some **Exception Module**. The analysis of the exception usually starts with that module's function.

## Origin Module

After looking at the **Exception Module**, let's move down the call chain of functions from **Stack Trace** or **Managed Stack Trace** to find out the possible **Origin Module** where the problem possibly could have originated.

## Thread Patterns

The large group of debugging analysis patterns is about execution threads. Stack traces are usually related to some thread. Most Python programs are single-threaded from a Python language source code perspective, but other threads may be created in the same Python process by either an operating system or third-party libraries.

---

[6]https://learn.microsoft.com/en-us/sysinternals/downloads/notmyfault

# Spiking Thread

The **Spiking Thread** analysis pattern deals with problem cases when there is some significant CPU consumption and you want to identify the relevant thread and its **Stack Trace**. **Paratext** may help here with additional operating system data.

# Active Thread

If you look at some random process and one of its threads, you will see it sleeping or waiting. This is because there are dozens (or hundreds on multiuser systems) of processes and hundreds (or thousands on multiuser systems) of threads (especially on Windows) and only a few CPUs for simultaneous thread execution. Therefore, the presence of an **Active Thread** is something unusual that may trigger further investigation. Such a thread may not be a **Spiking Thread** and may be a temporary activity.

# Blocked Thread

Most threads are not **Active Threads**, and this should be normal. However, some threads may get your attention as **Blocked Threads**, waiting for interprocess communication, synchronization locks, file, and socket I/O. You will see some examples in later chapters when you look at problems related to interfacing with an operating system.

# Blocking Module

This analysis pattern should belong to the module patterns category, but I mention it here because after identifying a **Blocked Thread** and looking at its (**Managed**) **Stack Trace**, you can find out the **Blocking Module** where the blocking code is located and even find out the **Origin Module** where all blocking behavior originated.

# Synchronization Patterns

This group of patterns is about threads waiting for synchronization objects such as locks, mutexes, events, and critical sections.

# Wait Chain

A **Blocked Thread** may wait for another thread or a synchronization object owned by some other thread, and that thread may also wait. This waiting activity constitutes the so-called **Wait Chain**.

# Deadlock

When a **Wait Chain** is cyclic, we call it a **Deadlock**. The simplest case is when two processes or threads are waiting for each other: thread #1 holds a lock and is waiting for another lock that is held by thread #2, which is waiting for the lock held by thread #1.

# Livelock

A **Livelock** is a sort of a **Deadlock**, but threads wait only periodically, perhaps due to small wait timeouts or just being **Active Threads**.

# Memory Consumption Patterns

Finally, let's look at memory consumption. There are different types of memory: process heap, virtual memory, physical memory, and kernel pools.

# Memory Leak

The **Memory Leak** analysis pattern deals with diagnosing the problem of gradually increasing memory consumption. Here periodic memory dumps, tracing, and logging may help to identify memory usage patterns such as **Stack Traces** of allocation, deallocation, and their relative ratio. You will debug a memory leak example in Chapter 5.

# Handle Leak

A handle (or descriptor or ID) is a numeric representation of some resource, be it a process or thread, file, or operating system resource. The code is supposed to close it after usage, but it doesn't do so due to some code defect. **Handle Leaks** are closely related to **Memory Leaks** because handles refer to structures in memory, and forgetting to close does not free or release the associated dynamic memory (usually in kernel space).

# Case Study

To illustrate a few debugging analysis patterns, execute the code in Listing 4-5 on Linux.

***Listing 4-5.*** A Script to Illustrate Some Debugging Analysis Patterns

```python
# spiking-thread.py

import time
import threading
import math

def thread_func(spike):
    foo(spike)

def main():
    threads: list[threading.Thread] = []
    threads.append(threading.Thread(target=thread_func, args=[False]))
    threads.append(threading.Thread(target=thread_func, args=[True]))
    threads.append(threading.Thread(target=thread_func, args=[False]))
    for thread in threads:
        thread.start()
    for thread in threads:
        thread.join()

def foo(spike):
    bar(spike)

def bar(spike):
    while True:
        if spike:
            math.sqrt(2);
        else:
            time.sleep(1)

if __name__ == "__main__":
    main()
```

Run the script and wait for a few minutes. When executing the top command, you'll see one **Spiking Thread** consuming a lot of CPU time:

```
~/Python-Book/Chapter 4$ ps
   PID TTY          TIME CMD
 162082 pts/0    00:00:00 bash
 163195 pts/0    00:52:51 python3
 164966 pts/0    00:00:00 ps

~/Python-Book/Chapter 4$ ps -T
   PID   SPID TTY         TIME CMD
 162082 162082 pts/0    00:00:00 bash
 163195 163195 pts/0    00:00:00 python3
 163195 163197 pts/0    00:00:00 python3
 163195 163198 pts/0    00:53:19 python3
 163195 163199 pts/0    00:00:00 python3
 165070 165070 pts/0    00:00:00 ps

~/Python-Book/Chapter 4$ top -H -p 163195
top - 14:20:55 up 14 days,  3:30,  0 users,  load average: 1.00, 1.04, 1.04
Threads:   4 total,  1 running,   3 sleeping,   0 stopped,   0 zombie
%Cpu(s): 50.7 us,  0.4 sy,  0.0 ni, 48.7 id,  0.0 wa,  0.0 hi,  0.2
si,  0.0 st
MiB Mem :  11940.2 total,   8176.1 free,    729.2 used,   3035.0 buff/cache
MiB Swap:      0.0 total,      0.0 free,      0.0 used.  10996.1 avail Mem

   PID USER     PR  NI  VIRT   RES    SHR S  %CPU  %MEM
   TIME+ COMMAND
 163198 ubuntu   20   0  236060 11084  7596 R  99.6   0.1
56:46.22 python3
 163195 ubuntu   20   0  236060 11084  7596 S   0.0   0.1
0:00.15 python3
 163197 ubuntu   20   0  236060 11084  7596 S   0.0   0.1
0:00.13 python3
 163199 ubuntu   20   0  236060 11084  7596 S   0.0   0.1
0:00.10 python3
```

Note that the **Spiking Thread** ID is 163198.

Now you can either attach GDB to the running process or save a process core memory dump using the gcore tool. Take the second approach. If you don't have these tools on your Linux box, you need to install them for this case study. On our Ubuntu AArch64 cloud instance, the following command installed both tools:

```
~/Python-Book/Chapter 4$ sudo apt install gdb
```

Then save a core dump for later memory dump analysis:

```
~/Python-Book/Chapter 4$ sudo gcore 163195
[New LWP 163197]
[New LWP 163198]
[New LWP 163199]
[Thread debugging using libthread_db enabled]
Using host libthread_db library "/lib/aarch64-linux-gnu/libthread_db.so.1".
0x0000ffffbe76b268 in do_futex_wait.constprop () from /lib/aarch64-linux-
gnu/libpthread.so.0
warning: target file /proc/163195/cmdline contained unexpected null
characters
Saved corefile core.163195
[Inferior 1 (process 163195) detached]

~/Python-Book/Chapter 4$ ls -l
total 232444
-rw-r--r-- 1 root    root    238014728 Jul  8 15:51 core.163195
-rw-rw-r-- 1 ubuntu ubuntu        667 Jul  8 13:08 spiking-thread.py
```

For the analysis of various managed patterns, you also need to install Python debugging packages (if you use the Red Hat distribution flavor, please check this article[7]):

```
~/Python-Book/Chapter 4$ sudo apt install python3-dbg
```

---

[7]https://developers.redhat.com/articles/2021/09/08/debugging-python-c-extensions-gdb

Now load the saved memory dump into GDB, specifying the Python executable as the source of symbols:

```
~/Python-Book/Chapter 4$ which python3
/usr/bin/python3
```

```
~/Python-Book/Chapter 4$ gdb -c core.163195 -se /usr/bin/python3
GNU gdb (Ubuntu 9.2-0ubuntu1~20.04.1) 9.2
Copyright (C) 2020 Free Software Foundation, Inc.
License GPLv3+: GNU GPL version 3 or later <http://gnu.org/licenses/
gpl.html>
This is free software: you are free to change and redistribute it.
There is NO WARRANTY, to the extent permitted by law.
Type "show copying" and "show warranty" for details.
This GDB was configured as "aarch64-linux-gnu".
Type "show configuration" for configuration details.
For bug reporting instructions, please see:
<http://www.gnu.org/software/gdb/bugs/>.
Find the GDB manual and other documentation resources online at:
    <http://www.gnu.org/software/gdb/documentation/>.
For help, type "help".
Type "apropos word" to search for commands related to "word"...
Reading symbols from /usr/bin/python3...
Reading symbols from /usr/lib/debug/.build-id/13/3e34cf1253c9541cfe6e927e
f3df6d9ea56f0e.debug...
[New LWP 163195]
[New LWP 163197]
[New LWP 163198]
[New LWP 163199]
[Thread debugging using libthread_db enabled]
Using host libthread_db library "/lib/aarch64-linux-gnu/libthread_db.so.1".
Core was generated by `python3'.
#0  futex_abstimed_wait_cancelable (private=0, abstime=0x0, clockid=0,
expected=0, futex_word=0xffffb8000b60)
    at ../sysdeps/nptl/futex-internal.h:320
```

```
320     ../sysdeps/nptl/futex-internal.h: No such file or directory.
[Current thread is 1 (Thread 0xffffbe91f010 (LWP 163195))]
(gdb)
```

To get the current thread **Stack Trace**, you can use the bt command (use an option to reduce the output of each frame function's arguments), which is also an example of Python **Runtime Thread** and **Source Stack Trace** because debugging symbolic information for the Python executable is also available:

```
(gdb) thread
[Current thread is 1 (Thread 0xffffbe91f010 (LWP 163195))]

(gdb) bt -frame-arguments none
#0  futex_abstimed_wait_cancelable (private=..., abstime=..., clockid=...,
expected=..., futex_word=...) at ../sysdeps/nptl/futex-internal.h:320
#1  do_futex_wait (sem=..., abstime=..., clockid=...) at sem_
waitcommon.c:112
#2  0x0000ffffbe76b39c in __new_sem_wait_slow (sem=..., abstime=...,
clockid=...) at sem_waitcommon.c:184
#3  0x0000ffffbe76b43c in __new_sem_wait (sem=...) at sem_wait.c:42
#4  0x00000000004d3150 in PyThread_acquire_lock_timed (lock=...,
microseconds=..., intr_flag=...) at ../Python/thread_pthread.h:471
#5  0x00000000004b78f4 in acquire_timed (timeout=..., lock=...) at ../
Modules/_threadmodule.c:63
#6  lock_PyThread_acquire_lock (self=..., args=..., kwds=...) at ../
Modules/_threadmodule.c:146
#7  0x000000000048be84 in method_vectorcall_VARARGS_KEYWORDS (func=...,
args=..., nargsf=..., kwnames=...) at ../Objects/descrobject.c:332
#8  0x00000000004fe444 in _PyObject_Vectorcall (kwnames=..., nargsf=...,
args=..., callable=...) at ../Include/cpython/abstract.h:127
#9  call_function (kwnames=..., oparg=..., pp_stack=..., tstate=...) at ../
Python/ceval.c:4963
#10 _PyEval_EvalFrameDefault (f=..., throwflag=...) at ../Python/
ceval.c:3486
#11 0x00000000004fc604 in PyEval_EvalFrameEx (throwflag=..., f=...) at ../
Python/ceval.c:741
```

```
#12 _PyEval_EvalCodeWithName (_co=..., globals=..., locals=..., args=...,
argcount=..., kwnames=..., kwargs=..., kwcount=..., kwstep=..., defs=...,
    defcount=..., kwdefs=..., closure=..., name=..., qualname=...) at ../
    Python/ceval.c:4298
#13 0x000000000059693c in _PyFunction_Vectorcall (func=..., stack=...,
nargsf=..., kwnames=...) at ../Objects/call.c:428
#14 0x00000000004fe444 in _PyObject_Vectorcall (kwnames=..., nargsf=...,
args=..., callable=...) at ../Include/cpython/abstract.h:127
#15 call_function (kwnames=..., oparg=..., pp_stack=..., tstate=...) at ../
Python/ceval.c:4963
#16 _PyEval_EvalFrameDefault (f=..., throwflag=...) at ../Python/
ceval.c:3486
#17 0x00000000004fc604 in PyEval_EvalFrameEx (throwflag=..., f=...) at ../
Python/ceval.c:741
#18 _PyEval_EvalCodeWithName (_co=..., globals=..., locals=..., args=...,
argcount=..., kwnames=..., kwargs=..., kwcount=..., kwstep=..., defs=...,
    defcount=..., kwdefs=..., closure=..., name=..., qualname=...) at ../
    Python/ceval.c:4298
#19 0x000000000059693c in _PyFunction_Vectorcall (func=..., stack=...,
nargsf=..., kwnames=...) at ../Objects/call.c:428
#20 0x00000000004fe444 in _PyObject_Vectorcall (kwnames=..., nargsf=...,
args=..., callable=...) at ../Include/cpython/abstract.h:127
#21 call_function (kwnames=..., oparg=..., pp_stack=..., tstate=...) at ../
Python/ceval.c:4963
#22 _PyEval_EvalFrameDefault (f=..., throwflag=...) at ../Python/
ceval.c:3486
#23 0x0000000000596748 in PyEval_EvalFrameEx (throwflag=..., f=...) at ../
Python/ceval.c:741
#24 function_code_fastcall (globals=..., nargs=..., args=..., co=...) at
../Objects/call.c:284
#25 _PyFunction_Vectorcall (func=..., stack=..., nargsf=..., kwnames=...)
at ../Objects/call.c:411
#26 0x00000000004fe31c in _PyObject_Vectorcall (kwnames=..., nargsf=...,
args=..., callable=...) at ../Include/cpython/abstract.h:127
```

#27 call_function (kwnames=..., oparg=..., pp_stack=..., tstate=...) at ../
Python/ceval.c:4963
#28 _PyEval_EvalFrameDefault (f=..., throwflag=...) at ../Python/
ceval.c:3500
#29 0x00000000004fc604 in PyEval_EvalFrameEx (throwflag=..., f=...) at ../
Python/ceval.c:741
--Type <RET> for more, q to quit, c to continue without paging--
#30 _PyEval_EvalCodeWithName (_co=..., globals=..., locals=..., args=...,
argcount=..., kwnames=..., kwargs=..., kwcount=..., kwstep=..., defs=...,
    defcount=..., kwdefs=..., closure=..., name=..., qualname=...) at ../
    Python/ceval.c:4298
#31 0x00000000006629f0 in PyEval_EvalCodeEx (closure=..., kwdefs=...,
defcount=..., defs=..., kwcount=..., kws=..., argcount=..., args=...,
    locals=..., globals=..., _co=...) at ../Python/ceval.c:4327
#32 PyEval_EvalCode (co=..., globals=..., locals=...) at ../Python/
ceval.c:718
#33 0x000000000064eac0 in run_eval_code_obj (co=..., globals=...,
locals=...) at ../Python/pythonrun.c:1166
#34 0x000000000064eb8c in run_mod (mod=..., filename=..., globals=...,
locals=..., flags=..., arena=...) at ../Python/pythonrun.c:1188
#35 0x000000000064ec84 in pyrun_file (fp=..., filename=..., start=...,
globals=..., locals=..., closeit=..., flags=...)
    at ../Python/pythonrun.c:1085
#36 0x000000000064f098 in pyrun_simple_file (flags=..., closeit=...,
filename=..., fp=...) at ../Python/pythonrun.c:439
#37 PyRun_SimpleFileExFlags (fp=..., filename=..., closeit=..., flags=...)
at ../Python/pythonrun.c:472
#38 0x0000000000069cdfc in pymain_run_file (cf=..., config=...) at ../
Modules/main.c:385
#39 pymain_run_python (exitcode=...) at ../Modules/main.c:610
#40 Py_RunMain () at ../Modules/main.c:689
#41 0x0000000000069d5d0 in pymain_main (args=...) at ../Modules/main.c:719
#42 0x0000000000069d624 in Py_BytesMain (argc=..., argv=...) at ../Modules/
main.c:743

```
#43 0x0000ffffbe7abe10 in __libc_start_main (main=..., argc=..., argv=...,
init=..., fini=..., rtld_fini=..., stack_end=...)
    at ../csu/libc-start.c:308
#44 0x000000000059bb68 in _start () at ../Objects/abstract.c:17
Backtrace stopped: previous frame identical to this frame (corrupt stack?)
```

To get the current thread's **Managed Stack Trace**, you can use the py-bt command:

```
(gdb) py-bt
Traceback (most recent call first):
  File "/usr/lib/python3.8/threading.py", line 1027, in _wait_for_
  tstate_lock
    elif lock.acquire(block, timeout):
  File "/usr/lib/python3.8/threading.py", line 1011, in join
    self._wait_for_tstate_lock()
  File "spiking-thread.py", line 20, in main
    thread.join()
  File "spiking-thread.py", line 33, in <module>
    main()
```

To check the available threads, you can use the info threads command:

```
(gdb) info threads
  Id   Target Id                            Frame
* 1    Thread 0xffffbe91f010 (LWP 163195) futex_abstimed_wait_
       cancelable (private=0, abstime=0x0, clockid=0, expected=0, futex_
       word=0xffffb8000b60)
    at ../sysdeps/nptl/futex-internal.h:320
  2    Thread 0xffffbe0a81e0 (LWP 163197) 0x0000ffffbe855620 in __GI__
       select (nfds=nfds@entry=0, readfds=readfds@entry=0x0,
    writefds=writefds@entry=0x0, exceptfds=exceptfds@entry=0x0,
    timeout=timeout@entry=0xffffbe0a6a80) at ../sysdeps/unix/sysv/linux/
    select.c:53
  3    Thread 0xffffbd8a71e0 (LWP 163198) 0x000000000057d758 in PyLong_
       AsDouble (v=2) at ../Objects/longobject.c:3043
  4    Thread 0xffffbd0a61e0 (LWP 163199) 0x0000ffffbe855620 in __GI__
       select (nfds=nfds@entry=0, readfds=readfds@entry=0x0,
```

```
writefds=writefds@entry=0x0, exceptfds=exceptfds@entry=0x0,
timeout=timeout@entry=0xffffbd0a4a80) at ../sysdeps/unix/sysv/linux/
select.c:53
```

You can see that the spiking thread 163198 from the **Paratext** output of the top command is thread #3. You can switch to it and examine the **Managed Stack Trace**:

```
(gdb) thread 3
[Switching to thread 3 (Thread 0xffffbd8a71e0 (LWP 163198))]
#0  0x000000000057d758 in PyLong_AsDouble (v=2) at ../Objects/
longobject.c:3043
3043    ../Objects/longobject.c: No such file or directory.

(gdb) py-bt
Traceback (most recent call first):
  <built-in method sqrt of module object at remote 0xffffbe1096d0>
  File "spiking-thread.py", line 28, in bar
    math.sqrt(2);
  File "spiking-thread.py", line 23, in foo
    bar(spike)
  File "spiking-thread.py", line 8, in thread_func
    foo(spike)
  File "/usr/lib/python3.8/threading.py", line 870, in run
    self._target(*self._args, **self._kwargs)
  File "/usr/lib/python3.8/threading.py", line 932, in _bootstrap_inner
    self.run()
  File "/usr/lib/python3.8/threading.py", line 890, in _bootstrap
    self._bootstrap_inner()
```

You also see this thread is an **Active Thread** since it is not waiting or sleeping. You can use the py-list command to show the Python source code and the py-up and py-down commands to navigate backtrace frames. You can also use the py-locals command to check function parameters and local variables for the current frame and the py-print command to check any global variable (you need to be in some Python frame to be able to do that).

```
(gdb) py-list
  23          bar(spike)
  24
  25      def bar(spike):
  26          while True:
  27              if spike:
 >28                  math.sqrt(2);
  29              else:
  30                  time.sleep(1)
  31
  32      if __name__ == "__main__":
  33          main()

(gdb) py-local
Unable to read information on python frame

(gdb) py-down
#10 Frame 0xffffbe1226c0, for file spiking-thread.py, line 28, in bar
(spike=True)
    math.sqrt(2);

(gdb) py-locals
spike = True

(gdb) py-print __file__
global '__file__' = 'spiking-thread.py'

(gdb) py-print __name__
global '__name__' = '__main__'
```

It is also good to check other threads to see if there are any anomalies. To see the (managed) **Stack Trace Collection**, you can use these commands (I removed some output for clarity):

```
(gdb) thread apply all bt
```

```
Thread 4 (Thread 0xffffbd0a61e0 (LWP 163199)):
#0  0x0000ffffbe855620 in __GI___select (nfds=nfds@entry=0, readfds=readfds@
entry=0x0, writefds=writefds@entry=0x0, exceptfds=exceptfds@entry=0x0,
timeout=timeout@entry=0xffffbd0a4a80) at ../sysdeps/unix/sysv/linux/select.c:53
```

#1   0x00000000005ce6dc in pysleep (secs=<optimized out>) at ../Modules/
timemodule.c:1866
#2   time_sleep (self=<optimized out>, obj=<optimized out>) at ../Modules/
timemodule.c:338
...

Thread 3 (Thread 0xffffbd8a71e0 (LWP 163198)):
#0   0x000000000057d758 in PyLong_AsDouble (v=2) at ../Objects/
longobject.c:3043
#1   0x0000000000572b6c in long_float (v=<optimized out>) at ../Objects/
longobject.c:5085
#2   0x000000000069177c in PyFloat_AsDouble (op=2) at ../Objects/
floatobject.c:263
#3   0x00000000004adc64 in PyFloat_AsDouble (op=2) at ../Objects/
floatobject.c:129
#4   math_1_to_whatever (can_overflow=0, from_double_func=<optimized out>,
func=<optimized out>, arg=2) at ../Modules/mathmodule.c:935
#5   math_1 (can_overflow=0, func=<optimized out>, arg=2) at ../Modules/
mathmodule.c:1013
#6   math_sqrt (self=<optimized out>, args=2) at ../Modules/
mathmodule.c:1209
...

Thread 2 (Thread 0xffffbe0a81e0 (LWP 163197)):
#0   0x0000ffffbe855620 in __GI___select (nfds=nfds@entry=0,
readfds=readfds@entry=0x0, writefds=writefds@entry=0x0,
exceptfds=exceptfds@entry=0x0, timeout=timeout@entry=0xffffbe0a6a80) at ../
sysdeps/unix/sysv/linux/select.c:53
#1   0x00000000005ce6dc in pysleep (secs=<optimized out>) at ../Modules/
timemodule.c:1866
#2   time_sleep (self=<optimized out>, obj=<optimized out>) at ../Modules/
timemodule.c:338
...
Thread 1 (Thread 0xffffbe91f010 (LWP 163195)):

```
#0  futex_abstimed_wait_cancelable (private=0, abstime=0x0, clockid=0,
expected=0, futex_word=0xffffb8000b60) at ../sysdeps/nptl/futex-
internal.h:320
#1  do_futex_wait (sem=sem@entry=0xffffb8000b60, abstime=0x0, clockid=0) at
sem_waitcommon.c:112
--Type <RET> for more, q to quit, c to continue without paging--
#2  0x0000ffffbe76b39c in __new_sem_wait_slow (sem=sem@
entry=0xffffb8000b60, abstime=0x0, clockid=0) at sem_waitcommon.c:184
```

(gdb) **thread apply all py-bt**

```
Thread 4 (Thread 0xffffbd0a61e0 (LWP 163199)):
Traceback (most recent call first):
  <built-in method sleep of module object at remote 0xffffbe282b80>
  File "spiking-thread.py", line 30, in bar
    time.sleep(1)
  File "spiking-thread.py", line 23, in foo
    bar(spike)
  File "spiking-thread.py", line 8, in thread_func
    foo(spike)
  File "/usr/lib/python3.8/threading.py", line 870, in run
    self._target(*self._args, **self._kwargs)
  File "/usr/lib/python3.8/threading.py", line 932, in _bootstrap_inner
    self.run()
  File "/usr/lib/python3.8/threading.py", line 890, in _bootstrap
    self._bootstrap_inner()

Thread 3 (Thread 0xffffbd8a71e0 (LWP 163198)):
Traceback (most recent call first):
  <built-in method sqrt of module object at remote 0xffffbe1096d0>
  File "spiking-thread.py", line 28, in bar
    math.sqrt(2);
  File "spiking-thread.py", line 23, in foo
    bar(spike)
  File "spiking-thread.py", line 8, in thread_func
    foo(spike)
  File "/usr/lib/python3.8/threading.py", line 870, in run
```

```
    self._target(*self._args, **self._kwargs)
  File "/usr/lib/python3.8/threading.py", line 932, in _bootstrap_inner
    self.run()
  File "/usr/lib/python3.8/threading.py", line 890, in _bootstrap
    self._bootstrap_inner()
--Type <RET> for more, q to quit, c to continue without paging--

Thread 2 (Thread 0xffffbe0a81e0 (LWP 163197)):
Traceback (most recent call first):
  <built-in method sleep of module object at remote 0xffffbe282b80>
  File "spiking-thread.py", line 30, in bar
    time.sleep(1)
  File "spiking-thread.py", line 23, in foo
    bar(spike)
  File "spiking-thread.py", line 8, in thread_func
    foo(spike)
  File "/usr/lib/python3.8/threading.py", line 870, in run
    self._target(*self._args, **self._kwargs)
  File "/usr/lib/python3.8/threading.py", line 932, in _bootstrap_inner
    self.run()
  File "/usr/lib/python3.8/threading.py", line 890, in _bootstrap
    self._bootstrap_inner()

Thread 1 (Thread 0xffffbe91f010 (LWP 163195)):
Traceback (most recent call first):
  File "/usr/lib/python3.8/threading.py", line 1027, in _wait_for_
tstate_lock
    elif lock.acquire(block, timeout):
  File "/usr/lib/python3.8/threading.py", line 1011, in join
    self._wait_for_tstate_lock()
  File "spiking-thread.py", line 20, in main
    thread.join()
  File "spiking-thread.py", line 33, in <module>
    main()
```

You can see, indeed, that all other threads are sleeping, not active.

You can do a postmortem debugging session since the dump file can be analyzed after the program finishes, but for live debugging, you can attach GDB to a running process and execute the same commands, for example:

```
(gdb) q
~/Python-Book/Chapter 4$ sudo gdb -p 163195 -se /usr/bin/python3
GNU gdb (Ubuntu 9.2-0ubuntu1~20.04.1) 9.2
Copyright (C) 2020 Free Software Foundation, Inc.
License GPLv3+: GNU GPL version 3 or later <http://gnu.org/licenses/
gpl.html>
This is free software: you are free to change and redistribute it.
There is NO WARRANTY, to the extent permitted by law.
Type "show copying" and "show warranty" for details.
This GDB was configured as "aarch64-linux-gnu".
Type "show configuration" for configuration details.
For bug reporting instructions, please see:
<http://www.gnu.org/software/gdb/bugs/>.
Find the GDB manual and other documentation resources online at:
    <http://www.gnu.org/software/gdb/documentation/>.

For help, type "help".
Type "apropos word" to search for commands related to "word"...
Reading symbols from /usr/bin/python3...
Reading symbols from /usr/lib/debug/.build-id/13/3e34cf1253c9541cfe6e927e
f3df6d9ea56f0e.debug...
Attaching to program: /usr/bin/python3, process 163195
[New LWP 163197]
[New LWP 163198]
[New LWP 163199]
[Thread debugging using libthread_db enabled]
Using host libthread_db library "/lib/aarch64-linux-gnu/libthread_db.so.1".
futex_abstimed_wait_cancelable (private=0, abstime=0x0, clockid=0,
expected=0, futex_word=0xffffb8000b60) at ../sysdeps/nptl/futex-
internal.h:320
320     ../sysdeps/nptl/futex-internal.h: No such file or directory.
```

```
(gdb) py-bt
Traceback (most recent call first):
  File "/usr/lib/python3.8/threading.py", line 1027, in _wait_for_
  tstate_lock
    elif lock.acquire(block, timeout):
  File "/usr/lib/python3.8/threading.py", line 1011, in join
    self._wait_for_tstate_lock()
  File "spiking-thread.py", line 20, in main
    thread.join()
  File "spiking-thread.py", line 33, in <module>
    main()
```

# Summary

In this chapter, you overviewed the most common debugging analysis patterns and practiced some of them in the context of a Python case study. You will look at more debugging analysis patterns in the chapter about Python interfacing with operating systems. The next chapter looks at debugging implementation patterns in the context of the Python command-line debugger, **pdb**.

# CHAPTER 5

# Debugging Implementation Patterns

In the previous chapter, I introduced specific debugging analysis patterns in the context of several Python case studies. In this chapter, you will look at specific debugging implementation patterns in the context of the Python debugger, **pdb**. (Figure 5-1).

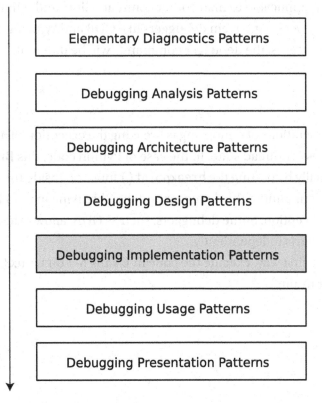

**Figure 5-1.** *Pattern-oriented debugging process and debugging implementation patterns*

© Dmitry Vostokov 2024
D. Vostokov, *Python Debugging for AI, Machine Learning, and Cloud Computing*,
https://doi.org/10.1007/978-1-4842-9745-2_5

I will skip architecture and design debugging patterns since, for simple cases, you may not need to think about them implicitly. Compare that to the implementation of a simple command-line utility or script: you don't usually use a heavy development process there explicitly. In such a case, architectural and design decisions are already implicit, for example, in the need to write a command-line utility. Also, when learning to program, it is better to start with some concrete useful implementation examples. Therefore, I will talk about debugging architecture and design patterns when I have enough case studies to generalize and when I see some non-trivial problems.

# Overview of Patterns

I now introduce debugging implementation pattern language in the context of **pdb**, the Python debugger[1]. This pattern language is not specific to this debugger; the same pattern language is applicable to many other command line[2] and IDE Python debuggers and even to native operating system debuggers like GDB and WinDbg. Of course, to implement these patterns, the debugger commands will be different.

# Break-Ins

One of the basic operations of debugging is breaking the execution of a process or thread to inspect its process or thread state. In the case of Python code, this **Break-in** can be done either internally by putting the breakpoint() function inside the code or externally via executing the script under the debugger and then breaking in via a keyboard interrupt (^C) or exception. Some debuggers, such as GDB, allow external **Break-ins** after the program starts independently.

To illustrate the first way, execute the code in Listing 5-1 on Linux (for Windows, you should get similar results).

---

[1] https://docs.python.org/3/library/pdb.html
[2] https://wiki.python.org/moin/PythonDebuggingTools

***Listing 5-1.*** A Simple Script to Test an Internal Break-in

```
# test-breakpoint.py
def main():
    foo()

def foo():
    bar()

def bar():
    while True:
        breakpoint()

if __name__ == "__main__":
    main()
```

After the **Break-in,** you can inspect the managed stack trace (where command or w) and resume execution (continue command or c). The quit command (or q) exits the debugger and aborts the script. There are other commands you will learn when you look at other debugging implementation patterns.

```
~/Chapter5$ python3 test-breakpoint.py
> ~/Chapter5/test-breakpoint.py(9)bar()
-> breakpoint()
(Pdb) w
  ~/Chapter5/test-breakpoint.py(12)<module>()
-> main()
  ~/Chapter5/test-breakpoint.py(2)main()
-> foo()
  ~/Chapter5/test-breakpoint.py(5)foo()
-> bar()
> ~/Chapter5/test-breakpoint.py(9)bar()
-> breakpoint()
(Pdb) c
> ~/Chapter5/test-breakpoint.py(9)bar()
-> breakpoint()
(Pdb) q
```

```
Traceback (most recent call last):
  File "test-breakpoint.py", line 12, in <module>
    main()
  File "test-breakpoint.py", line 2, in main
    foo()
  File "test-breakpoint.py", line 5, in foo
    bar()
  File "test-breakpoint.py", line 9, in bar
    breakpoint()
  File "test-breakpoint.py", line 9, in bar
    breakpoint()
  File "/usr/lib/python3.7/bdb.py", line 88, in trace_dispatch
    return self.dispatch_line(frame)
  File "/usr/lib/python3.7/bdb.py", line 113, in dispatch_line
    if self.quitting: raise BdbQuit
bdb.BdbQuit
~/Chapter5$
```

For the external **Break-in**, you specify the *pdb* module when you execute the script. To illustrate this way, execute the code in Listing 5-2 on Linux (for Windows, you should get similar results).

***Listing 5-2.*** A Simple Script to Test an External Break-In and Other Patterns

```
# test-pdb.py
import time

class cls:
    def __init__(self, field1, field2):
        self.field1 = field1
        self.field2 = field2

def main():
    func = "main"
    while True:
        foo()
        time.sleep(1)
```

```
def foo():
    func = "foo"
    bar("argument")

def bar(arg):
    func = "bar"
    obj = [1, 2]
    obj = cls(1, 2)

if __name__ == "__main__":
    main()
```

The debugger stops before executing the first script line. You can continue execution and do **Break-in** at any time using ^C (or by sending the SIGINT signal using the kill command from a separate terminal window).

```
~/Chapter5$ python3 -m pdb test-pdb.py
> ~/Chapter5/test-pdb.py(1)<module>()
-> import time
(Pdb) c
^C
Program interrupted. (Use 'cont' to resume).
> ~/Chapter5/test-pdb.py(11)main()
-> foo()
(Pdb) w
  /usr/lib/python3.7/runpy.py(193)_run_module_as_main()
-> "__main__", mod_spec)
  /usr/lib/python3.7/runpy.py(85)_run_code()
-> exec(code, run_globals)
  /usr/lib/python3.7/pdb.py(1728)<module>()
-> pdb.main()
  /usr/lib/python3.7/pdb.py(1701)main()
-> pdb._runscript(mainpyfile)
  /usr/lib/python3.7/pdb.py(1570)_runscript()
-> self.run(statement)
  /usr/lib/python3.7/bdb.py(585)run()
-> exec(cmd, globals, locals)
```

```
  <string>(1)<module>()
  ~/Chapter5/test-pdb.py(1)<module>()
-> import time
> ~/Chapter5/test-pdb.py(11)main()
-> foo()
(Pdb)
```

Let's continue exploring other pdb commands in the pattern examples below.

## Code Breakpoint

You don't need to pollute your code with breakpoint() function calls each time you want to do **Break-in** at different locations in the script. After the initial **Break-in**, you can use the break (b) command to put **Code Breakpoint** at filename:lineno or a function name. For example, let's put such a breakpoint whenever the bar function is called:

```
(Pdb) b bar
Breakpoint 1 at ~/Chapter5/test-pdb.py:18
(Pdb) c
> ~/Chapter5/test-pdb.py(19)bar()
-> func = "bar"
(Pdb) w
  /usr/lib/python3.7/runpy.py(193)_run_module_as_main()
-> "__main__", mod_spec)
  /usr/lib/python3.7/runpy.py(85)_run_code()
-> exec(code, run_globals)
  /usr/lib/python3.7/pdb.py(1728)<module>()
-> pdb.main()
  /usr/lib/python3.7/pdb.py(1701)main()
-> pdb._runscript(mainpyfile)
  /usr/lib/python3.7/pdb.py(1570)_runscript()
-> self.run(statement)
  /usr/lib/python3.7/bdb.py(585)run()
-> exec(cmd, globals, locals)
  <string>(1)<module>()
  ~/Chapter5/test-pdb.py(1)<module>()
```

```
-> import time
  ~/Chapter5/test-pdb.py(11)main()
-> foo()
  ~/Chapter5/test-pdb.py(16)foo()
-> bar("argument")
> ~/Chapter5/test-pdb.py(19)bar()
-> func = "bar"
```

The arrow points to the next line to be executed. The breakpoint command without arguments lists available breakpoints, their numbers, and their information.

```
(Pdb) b
Num Type         Disp Enb   Where
1   breakpoint   keep yes   at ~/Chapter5/test-pdb.py:18
        breakpoint already hit 1 time
```

Then you can continue until any of the breakpoints are hit again. The clear (cl) command removes the existing breakpoint numbers. If you need a breakpoint that is executed only once, use the tbreak command instead of break. You can also enable and disable breakpoints by their numbers. For high-frequency breakpoints, you can also specify conditions to trigger. I provide such an example in the case study at the end of this chapter.

## Code Trace

Once you are inside some function, either via a **Break-in** or **Code Breakpoint**, you can start executing lines one by one using the step (s) command. If you want to treat the function call as one line and not go inside to execute line by line, you should use the next (n) command. If you accidentally step into a function, you can continue using the return (r) command. If you want to skip tracing some code lines, you can use the until (unt) command and specify the line number. If you even want to skip their execution, you can use the jump (j) command and specify the target line number.

```
(Pdb) c
> ~/Chapter5/test-pdb.py(19)bar()
-> func = "bar"
(Pdb) s
```

71

```
> ~/Chapter5/test-pdb.py(20)bar()
-> obj = [1, 2]
(Pdb) s
> ~/Chapter5/test-pdb.py(21)bar()
-> obj = cls(1, 2)
(Pdb) s
--Call--
> ~/Chapter5/test-pdb.py(4)__init__()
-> def __init__(self, field1, field2):
(Pdb) r
--Return--
> ~/Chapter5/test-pdb.py(6)__init__()->None
-> self.field2 = field2
(Pdb) r
--Return--
> ~/Chapter5/test-pdb.py(21)bar()->None
-> obj = cls(1, 2)
(Pdb) s
--Return--
> ~/Chapter5/test-pdb.py(16)foo()->None
-> bar("argument")
(Pdb) s
> ~/Chapter5/test-pdb.py(12)main()
-> time.sleep(1)
(Pdb) c
> ~/Chapter5/test-pdb.py(19)bar()
-> func = "bar"
(Pdb) n
> ~/Chapter5/test-pdb.py(20)bar()
-> obj = [1, 2]
(Pdb) n
> ~/Chapter5/test-pdb.py(21)bar()
-> obj = cls(1, 2)
(Pdb) n
```

```
--Return--
> ~/Chapter5/test-pdb.py(21)bar()->None
-> obj = cls(1, 2)
(Pdb) n
--Return--
> ~/Chapter5/test-pdb.py(16)foo()->None
-> bar("argument")
(Pdb) n
> ~/Chapter5/test-pdb.py(12)main()
-> time.sleep(1)
(Pdb)
```

# Scope

When you do **Break-in**, you are in a particular execution **Scope** that includes the currently executing function, its arguments, already defined local variables, and variables accessible from outer and global contexts. You can change the scope using the up (u) and down (d) commands to navigate up and down in the traceback of called functions. To see the current code location, you can use the list (l) command.

```
(Pdb) c
> ~/Chapter5/test-pdb.py(19)bar()
-> func = "bar"
(Pdb) l
 14     def foo():
 15         func = "foo"
 16         bar("argument")
 17
 18 B   def bar(arg):
 19  ->     func = "bar"
 20         obj = [1, 2]
 21         obj = cls(1, 2)
 22
 23     if __name__ == "__main__":
 24         main()
```

```
(Pdb) n
> ~/Chapter5/test-pdb.py(20)bar()
-> obj = [1, 2]
(Pdb) n
> ~/Chapter5/test-pdb.py(21)bar()
-> obj = cls(1, 2)
(Pdb) l
 16            bar("argument")
 17
 18 B   def bar(arg):
 19           func = "bar"
 20           obj = [1, 2]
 21  ->       obj = cls(1, 2)
 22
 23      if __name__ == "__main__":
 24          main()
[EOF]
(Pdb) a
arg = 'argument'
(Pdb) locals()
{'arg': 'argument', 'func': 'bar', 'obj': [1, 2]}
(Pdb) u
> ~/Chapter5/test-pdb.py(16)foo()
-> bar("argument")
(Pdb) l
 11               foo()
 12               time.sleep(1)
 13
 14      def foo():
 15           func = "foo"
 16  ->       bar("argument")
 17
 18 B   def bar(arg):
 19           func = "bar"
 20           obj = [1, 2]     .
```

```
 21             obj = cls(1, 2)
(Pdb) locals()
{'func': 'foo'}
(Pdb) u
> ~/Chapter5/test-pdb.py(11)main()
-> foo()
(Pdb) l
  6                 self.field2 = field2
  7
  8       def main():
  9           func = "main"
 10           while True:
 11   ->          foo()
 12               time.sleep(1)
 13
 14       def foo():
 15           func = "foo"
 16           bar("argument")
(Pdb) locals()
{'func': 'main'}
(Pdb)
```

# Variable Value

Obviously, when debugging, you are interested in **Variable Values**. Here you can use the p and pp (pretty-print) commands. You can also use the usual Python functions for this.

```
(Pdb) c
> ~/Chapter5/test-pdb.py(19)bar()
-> func = "bar"
(Pdb) n
> ~/Chapter5/test-pdb.py(20)bar()
-> obj = [1, 2]
(Pdb) n
> ~/Chapter5/test-pdb.py(21)bar()
-> obj = cls(1, 2)
```

```
(Pdb) p obj
[1, 2]
(Pdb) n
--Return--
> ~/Chapter5/test-pdb.py(21)bar()->None
-> obj = cls(1, 2)
(Pdb) p obj
<__main__.cls object at 0x7f7feb5b1e48>
(Pdb) vars(obj)
{'field1': 1, 'field2': 2}
(Pdb)
```

# Type Structure

When seeing an unfamiliar object, you may want to examine its **Type Structure**. Here you can use the appropriate Python modules and functions for exploration:

```
(Pdb) type(obj)
<class '__main__.cls'>
(Pdb) dir(obj)
['__class__', '__delattr__', '__dict__', '__dir__', '__doc__', '__eq__', '__format__', '__ge__', '__getattribute__', '__gt__', '__hash__', '__init__', '__init_subclass__', '__le__', '__lt__', '__module__', '__ne__', '__new__', '__reduce__', '__reduce_ex__', '__repr__', '__setattr__', '__sizeof__', '__str__', '__subclasshook__', '__weakref__', 'field1', 'field2']
(Pdb) vars(cls)
Mapping ({'__module__': '__main__', '__init__': <function cls.__init__ at 0x7f7feb5d4f28>, '__dict__': <attribute '__dict__' of 'cls' objects>, '__weakref__': <attribute '__weakref__' of 'cls' objects>, '__doc__': None})
(Pdb) import pprint
(Pdb) pprint.pprint(vars(cls))
mappingproxy({'__dict__': <attribute '__dict__' of 'cls' objects>,
              '__doc__': None,
              '__init__': <function cls.__init__ at 0x7f7feb5d4f28>,
              '__module__': '__main__',
              '__weakref__': <attribute '__weakref__' of 'cls' objects>})
```

```
(Pdb) pprint.pprint(dir(cls))
['__class__',
 '__delattr__',
 '__dict__',
 '__dir__',
 '__doc__',
 '__eq__',
 '__format__',
 '__ge__',
 '__getattribute__',
 '__gt__',
 '__hash__',
 '__init__',
 '__init_subclass__',
 '__le__',
 '__lt__',
 '__module__',
 '__ne__',
 '__new__',
 '__reduce__',
 '__reduce_ex__',
 '__repr__',
 '__setattr__',
 '__sizeof__',
 '__str__',
 '__subclasshook__',
 '__weakref__']
(Pdb) pprint.pprint(dir(obj))
['__class__',
 '__delattr__',
 '__dict__',
 '__dir__',
 '__doc__',
 '__eq__',
 '__format__',
```

```
 '__ge__',
 '__getattribute__',
 '__gt__',
 '__hash__',
 '__init__',
 '__init_subclass__',
 '__le__',
 '__lt__',
 '__module__',
 '__ne__',
 '__new__',
 '__reduce__',
 '__reduce_ex__',
 '__repr__',
 '__setattr__',
 '__sizeof__',
 '__str__',
 '__subclasshook__',
 '__weakref__',
 'field1',
 'field2']
(Pdb)
```

# Breakpoint Action

During debugging sessions, you want to automate **Code Tracing** and associated commands, such as printing tracebacks and **Variable Values**. Here **Breakpoint Actions** help by associating debugger commands with **Code Breakpoints**. For example, you can print a traceback each time you hit the breakpoint and automatically continue (use ^C to do **Break-in** again):

```
(Pdb) b
Num Type         Disp Enb   Where
1   breakpoint   keep yes   at ~/Chapter5/test-pdb.py:18
        breakpoint already hit 39 times
(Pdb) commands 1
```

```
(com) print("New hit!")
(com) w
(com) c
(Pdb) c
New hit!
  /usr/lib/python3.7/runpy.py(193)_run_module_as_main()
-> "__main__", mod_spec)
  /usr/lib/python3.7/runpy.py(85)_run_code()
-> exec(code, run_globals)
  /usr/lib/python3.7/pdb.py(1728)<module>()
-> pdb.main()
  /usr/lib/python3.7/pdb.py(1701)main()
-> pdb._runscript(mainpyfile)
  /usr/lib/python3.7/pdb.py(1570)_runscript()
-> self.run(statement)
  /usr/lib/python3.7/bdb.py(585)run()
-> exec(cmd, globals, locals)
  <string>(1)<module>()
  ~/Chapter5/test-pdb.py(1)<module>()
-> import time
  ~/Chapter5/test-pdb.py(11)main()
-> foo()
  ~/Chapter5/test-pdb.py(16)foo()
-> bar("argument")
> ~/Chapter5/test-pdb.py(19)bar()
-> func = "bar"
> ~/Chapter5/test-pdb.py(19)bar()
-> func = "bar"
New hit!
  /usr/lib/python3.7/runpy.py(193)_run_module_as_main()
-> "__main__", mod_spec)
  /usr/lib/python3.7/runpy.py(85)_run_code()
-> exec(code, run_globals)
  /usr/lib/python3.7/pdb.py(1728)<module>()
-> pdb.main()
```

```
  /usr/lib/python3.7/pdb.py(1701)main()
-> pdb._runscript(mainpyfile)
  /usr/lib/python3.7/pdb.py(1570)_runscript()
-> self.run(statement)
  /usr/lib/python3.7/bdb.py(585)run()
-> exec(cmd, globals, locals)
  <string>(1)<module>()
  ~/Chapter5/test-pdb.py(1)<module>()
-> import time
  ~/Chapter5/test-pdb.py(11)main()
-> foo()
  ~/Chapter5/test-pdb.py(16)foo()
-> bar("argument")
> ~/Chapter5/test-pdb.py(19)bar()
-> func = "bar"
> ~/Chapter5/test-pdb.py(19)bar()
-> func = "bar"
^C
Program interrupted. (Use 'cont' to resume).
--Call--
> /usr/lib/python3.7/bdb.py(319)set_trace()
-> def set_trace(self, frame=None):

(Pdb)
```

To clear commands for a breakpoint, just specify only the end command:

```
(Pdb) commands 1
(com) end
(Pdb) c
> /usr/lib/python3.7/bdb.py(279)_set_stopinfo()
-> self.returnframe = returnframe
(Pdb) c
> /usr/lib/python3.7/bdb.py(332)set_trace()
-> sys.settrace(self.trace_dispatch)
(Pdb) c
> ~/Chapter5/test-pdb.py(19)bar()
```

```
-> func = "bar"
(Pdb) c
> ~/Chapter5/test-pdb.py(19)bar()
-> func = "bar"
(Pdb)
```

Such automation may be useful in the remote debugging of cloud infrastructure and machine learning pipeline issues.

## Usage Trace

Often, for resource leak cases, you are interested in the **Usage Trace** of some object or container. Later, I will discuss yet another debugging implementation pattern called **Data Breakpoints** in the chapters devoted to native debugging. For now, I can suggest **Code Breakpoints** for places where objects of interest are used or on their accessor methods. Alternatively, you can use tracing and logging, either directly or by using decorators.

# Case Study

This case study models a problem I observed with a cloud host monitoring script. I keep the modeling code to a bare minimum to illustrate the usage of debugging implementation patterns.

## Elementary Diagnostics Patterns

There is a script that runs for long periods of time and monitors process creation. After some time, the counter value for memory usage starts continuously increasing.

## Debugging Analysis Patterns

The script is very small and uses a dictionary to keep previously existing process information for some time before removing it. A **Memory Leak (Process Heap)** is the obvious hypothesis here. For more complex scripts and programs, especially those that use third-party modules, some real memory analysis is required to differentiate between different types of leaks: process heap, virtual memory, and handles.

# Debugging Implementation Patterns

To debug, you do **Break-in** externally by starting the `process-monitoring.py` script from Listing 5-3 under pdb.

***Listing 5-3.*** Scripts to Model the Process Monitoring Case Study

```python
# process-monitoring.py
import processes
import filemon
import time

def main():
    procs = processes.Processes()
    files = filemon.Files()
    for pid in range (1, 10):
        procs.add_process(pid, "info")
    while True:
        pid += 1
        procs.add_process(pid, "info")
        time.sleep(1)
        procs.remove_process(pid)
        time.sleep(1)
        files.process_files()

if __name__ == "__main__":
    main()

# processes.py
class Processes:
    _singleton = None

    @staticmethod
    def __new__(cls):
        self = Processes._singleton
        if not self:
            Processes._singleton = self = super().__new__(cls)
        return self
```

```
    def __init__(self):
        pass

    _procinfo = {}

    def add_process(self, pid, info):
        Processes._procinfo[pid] = info

    def remove_process(self, pid):
        del Processes._procinfo[pid]

# filemon.py
import processes

class Files:
    def __init__(self):
        self._processes = processes.Processes()
        self._count = 0

    def process_files(self):
        self._count += 1
        if self._count > 25:
            self._processes.add_process(self._count, "")
```

```
~/Chapter5$ python3 -m pdb process-monitoring.py
> ~/Chapter5/process-monitoring.py(1)<module>()
-> import processes
(Pdb)
```

You continue execution for 10-15 seconds and then do **Break-in** again. You then inspect the **Variable Value** of the size of the Processes._procinfo dictionary.

```
(Pdb) c
^C
Program interrupted. (Use 'cont' to resume).
> ~/Chapter5/process-monitoring.py(14)main()
-> procs.remove_process(pid)
(Pdb) p len(processes.Processes._procinfo)
9
```

Since you know that the size of the dictionary starts increasing only after some time, you put a temporary **Code Breakpoint** on the add_process function to be hit only when the dictionary size exceeds 100. You continue execution again and wait.

```
(Pdb) tbreak processes.Processes.add_process, len(Processes._
procinfo) > 100
Breakpoint 1 at ~/Chapter5/processes.py:16
(Pdb) b
Num Type         Disp Enb   Where
1   breakpoint   del  yes   at ~/Chapter5/processes.py:16
        stop only if len(Processes._procinfo) > 100
(Pdb) c
Deleted breakpoint 1 at ~/Chapter5/processes.py:16
> ~/Chapter5/processes.py(17)add_process()
-> Processes._procinfo[pid] = info
(Pdb) b
(Pdb) p len(Processes._procinfo)
101
```

You now create two **Code Breakpoints** for the add_process and remove_process functions with a **Breakpoint Action** to print tracebacks of callers and continue.

```
(Pdb) b Processes.add_process
Breakpoint 2 at ~/Chapter5/processes.py:16
(Pdb) b Processes.remove_process
Breakpoint 3 at ~/Chapter5/processes.py:19
(Pdb) b
Num Type         Disp Enb   Where
2   breakpoint   keep yes   at ~/Chapter5/processes.py:16
3   breakpoint   keep yes   at ~/Chapter5/processes.py:19
(Pdb) commands 2
(com) print("ADD")
(com) w
(com) c
(Pdb) commands 3
(com) print("REMOVE")
```

```
(com) w
(com) c
(Pdb)
```

You continue execution again to gather some **Usage Traces** (for clarity, I omitted some frames from tracebacks).

```
(Pdb) c
REMOVE
...
-> import processes
  ~/Chapter5/process-monitoring.py(14)main()
-> procs.remove_process(pid)
> ~/Chapter5/processes.py(20)remove_process()
-> del Processes._procinfo[pid]
> ~/Chapter5/processes.py(20)remove_process()
-> del Processes._procinfo[pid]
ADD
...
-> import processes
  ~/Chapter5/process-monitoring.py(16)main()
-> files.process_files()
  ~/Chapter5/filemon.py(11)process_files()
-> self._processes.add_process(self._count, "")
> ~/Chapter5/processes.py(17)add_process()
-> Processes._procinfo[pid] = info
> ~/Chapter5/processes.py(17)add_process()
-> Processes._procinfo[pid] = info
ADD
...
-> import processes
  ~/Chapter5/process-monitoring.py(12)main()
-> procs.add_process(pid, "info")
> ~/Chapter5/processes.py(17)add_process()
-> Processes._procinfo[pid] = info
> ~/Chapter5/processes.py(17)add_process()
-> Processes._procinfo[pid] = info
```

**REMOVE**

...

```
-> import processes
  ~/Chapter5/process-monitoring.py(14)main()
-> procs.remove_process(pid)
> ~/Chapter5/processes.py(20)remove_process()
-> del Processes._procinfo[pid]
> ~/Chapter5/processes.py(20)remove_process()
-> del Processes._procinfo[pid]
```

**ADD**

...

```
-> import processes
  ~/Chapter5/process-monitoring.py(16)main()
```
**-> files.process_files()**
  **~/Chapter5/filemon.py(11)process_files()**
```
-> self._processes.add_process(self._count, "")
> ~/Chapter5/processes.py(17)add_process()
-> Processes._procinfo[pid] = info
> ~/Chapter5/processes.py(17)add_process()
-> Processes._procinfo[pid] = info
```

**ADD**

...

```
-> import processes
  ~/Chapter5/process-monitoring.py(12)main()
-> procs.add_process(pid, "info")
> ~/Chapter5/processes.py(17)add_process()
-> Processes._procinfo[pid] = info
> ~/Chapter5/processes.py(17)add_process()
-> Processes._procinfo[pid] = info
```

**REMOVE**

...

```
-> import processes
  ~/Chapter5/process-monitoring.py(14)main()
-> procs.remove_process(pid)
> ~/Chapter5/processes.py(20)remove_process()
```

```
-> del Processes._procinfo[pid]
> ~/Chapter5/processes.py(20)remove_process()
-> del Processes._procinfo[pid]
```

**ADD**

```
...
-> import processes
  ~/Chapter5/process-monitoring.py(16)main()
```

**-> files.process_files()**
**  ~/Chapter5/filemon.py(11)process_files()**

```
-> self._processes.add_process(self._count, "")
> ~/Chapter5/processes.py(17)add_process()
-> Processes._procinfo[pid] = info
> ~/Chapter5/processes.py(17)add_process()
-> Processes._procinfo[pid] = info
```

**ADD**

```
...
-> import processes
  ~/Chapter5/process-monitoring.py(12)main()
-> procs.add_process(pid, "info")
> ~/Chapter5/processes.py(17)add_process()
-> Processes._procinfo[pid] = info
> ~/Chapter5/processes.py(17)add_process()
-> Processes._procinfo[pid] = info
```

**REMOVE**

```
...
-> import processes
  ~/Chapter5/process-monitoring.py(14)main()
-> procs.remove_process(pid)
> ~/Chapter5/processes.py(20)remove_process()
-> del Processes._procinfo[pid]
> ~/Chapter5/processes.py(20)remove_process()
-> del Processes._procinfo[pid]
```

**ADD**

```
...
-> import processes
  ~/Chapter5/process-monitoring.py(16)main()
```

**-> files.process_files()**
  **~/Chapter5/filemon.py(11)process_files()**
-> self._processes.add_process(self._count, "")
> ~/Chapter5/processes.py(17)add_process()
-> Processes._procinfo[pid] = info
> ~/Chapter5/processes.py(17)add_process()
-> Processes._procinfo[pid] = info
**ADD**

...

-> import processes
  ~/Chapter5/process-monitoring.py(12)main()
-> procs.add_process(pid, "info")
> ~/Chapter5/processes.py(17)add_process()
-> Processes._procinfo[pid] = info
> ~/Chapter5/processes.py(17)add_process()
-> Processes._procinfo[pid] = info
**REMOVE**

...

-> import processes
  ~/Chapter5/process-monitoring.py(14)main()
-> procs.remove_process(pid)
> ~/Chapter5/processes.py(20)remove_process()
-> del Processes._procinfo[pid]
> ~/Chapter5/processes.py(20)remove_process()
-> del Processes._procinfo[pid]
**ADD**

...

-> import processes
  ~/Chapter5/process-monitoring.py(16)main()
**-> files.process_files()**
  **~/Chapter5/filemon.py(11)process_files()**
-> self._processes.add_process(self._count, "")
> ~/Chapter5/processes.py(17)add_process()
-> Processes._procinfo[pid] = info
> ~/Chapter5/processes.py(17)add_process()
-> Processes._procinfo[pid] = info

**ADD**

```
...
-> import processes
  ~/Chapter5/process-monitoring.py(12)main()
-> procs.add_process(pid, "info")
> ~/Chapter5/processes.py(17)add_process()
-> Processes._procinfo[pid] = info
> ~/Chapter5/processes.py(17)add_process()
-> Processes._procinfo[pid] = info
^C
Program interrupted. (Use 'cont' to resume).
--Call--
> /usr/lib/python3.7/bdb.py(319)set_trace()
-> def set_trace(self, frame=None):
(Pdb) q
```

You can see the number of ADD actions is twice as high as the REMOVE ones. Also, the ADD actions from the filemon.py module do not have any compensating REMOVE actions. This explains the **Memory Leak** debugging analysis pattern and suggests that changes in the file monitoring part are needed.

# Summary

In this chapter, you looked at debugging implementation patterns in the context of the Python command-line debugger. The next chapter looks at the same patterns in the context of various Python IDEs.

# CHAPTER 6

# IDE Debugging in the Cloud

This chapter will look at the debugging implementation patterns in the context of some Python IDEs.

## Visual Studio Code

Let's review the debugging implementation pattern language and case study from the previous chapter in the context of the Visual Studio Code IDE. I assume that you already have VS Code installed on your local machine with the required extensions. For the necessary steps, please look at the Python extension for Visual Studio Code README[1].

## WSL Setup

When preparing some case studies for this book, I used VS Code for editing and WSL (Windows Subsystem for Linux) running Debian to execute Python code (thus simulating a remote cloud instance). So, before looking at cloud setup, I want to provide you with instructions on how to make the same environment to debug locally using an IDE if you don't have access to a cloud environment. Please refer to the article[2] at `https://github.com/Microsoft/vscode-python` for how to prepare

---

[1] `https://github.com/Microsoft/vscode-python`

[2] `https://learn.microsoft.com/en-us/windows/wsl/tutorials/wsl-vscode`

© Dmitry Vostokov 2024

D. Vostokov, *Python Debugging for AI, Machine Learning, and Cloud Computing*,
https://doi.org/10.1007/978-1-4842-9745-2_6

such an environment. However, the recommended setup there might require some troubleshooting. On our Windows 11 WSL2 system, I had to add the following line to `.bashrc`:

```
alias code= "'/mnt/c/Users/[USER]/AppData/Local/Programs/Microsoft VS Code/
Code.exe'"
```

The required setup also installs the **Remote – SSH** VS Code extension necessary for the cloud setup.

## Cloud SSH Setup

To prepare a cloud remote debugging environment, I recommend the prerequisite steps in the article about remote development over SSH[3].

Below I provide an example of VS Code running on Windows 11 and connecting to our Ubuntu AArch64 system running in a cloud different from Azure, which was described in the referenced article. Since I already have the SSH key pair generated for us when I provisioned a cloud instance, I use it to connect.

The bottom-left corner of VS Code window has a connect button (Figure 6-1).

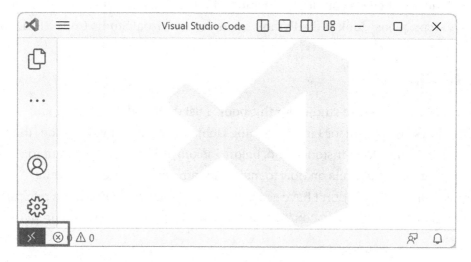

***Figure 6-1.***  *Open a remote window and use the connect button*

---

[3] `https://code.visualstudio.com/docs/remote/ssh-tutorial`

When you click it, you will see different options. Choose *Connect Current Window to Host* in the *Remote-SSH* section (Figure 6-2).

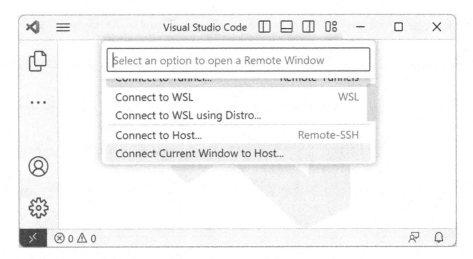

***Figure 6-2.*** *Options to connect to a remote host via SSH*

Then choose *Configure SSH Hosts* (Figure 6-3).

***Figure 6-3.*** *Options to configure remote hosts via SSH*

Next, choose a config file to write a cloud instance IP address, cloud instance username, and private key location (in our case, on Windows), as shown in Figure 6-4.

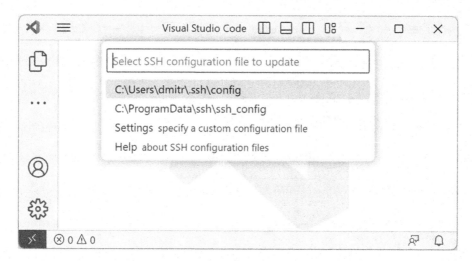

**Figure 6-4.** *Choosing a configuration file*

Choose a configuration file located in the C:\Users folder and add the required information (Figure 6-5).

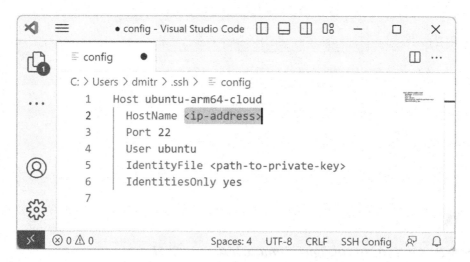

**Figure 6-5.** *A sample SSH configuration file. You need to specify an IP address of a cloud instance, username, and private key path*

Save the config file and click again the bottom-left *Open a Remote Window* button (Figure 6-2).

Then choose the host saved in the config file (Figure 6-6).

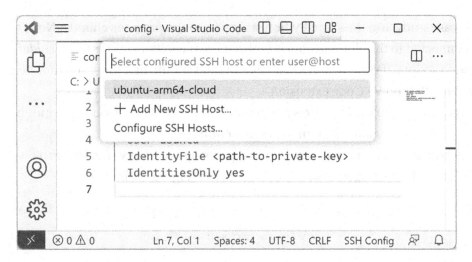

***Figure 6-6.*** *Choosing the host to connect*

You will get connected. The host is shown in the bottom-left connection button (Figure 6-7).

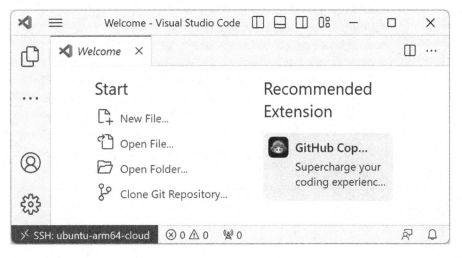

***Figure 6-7.*** *Connected host via SSH*

# Case Study

For this case study, use the same Python scripts from the previous chapter. Now, after being connected to the cloud instance, you can open a folder (*File ➤ Open Folder*) with the Python scripts. If you open or create Python files for the first time, you may be asked to install the Python VS Code extension.

To recall, the case study models a problem I observed with a cloud host monitoring script that, after time, starts leaking process memory.

To debug, start the `process-monitoring.py` using the F5 key (*Run ➤ Start Debugging*) and selecting *Python File* configuration (Figure 6-8).

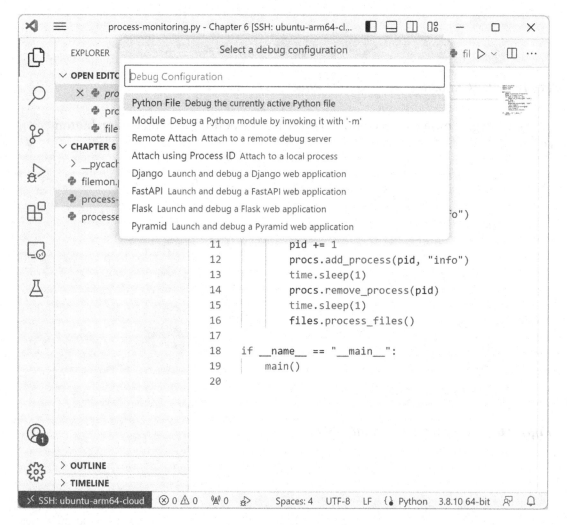

***Figure 6-8.*** *Choosing the debugging configuration*

The program is now running, and you do **Break-in** by using the *Pause* (F6) button on the *Debugging* floating toolbar (Figure 6-9).

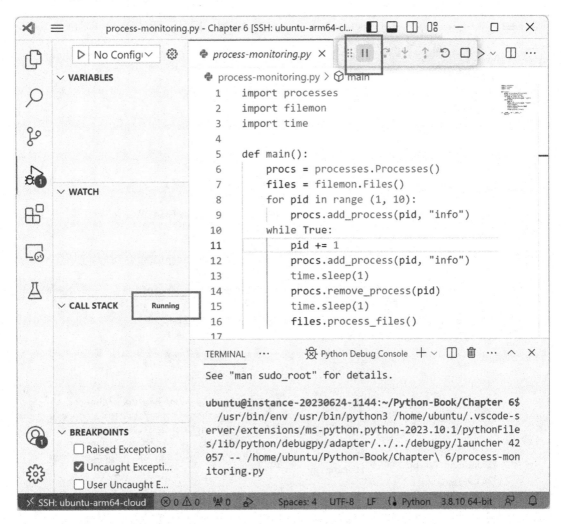

***Figure 6-9.*** *Running program and Debugging toolbar*

After a pause, choose the appropriate **Scope** in the call stack to make various variables accessible (Figure 6-10).

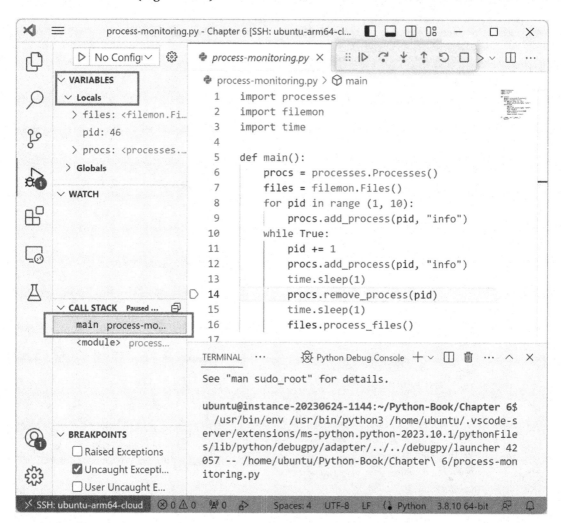

***Figure 6-10.*** *Debugging scope*

Now switch to the *Debug Console* tab and enter the expression to inspect the **Variable Value** of the size of the `Processes._procinfo` dictionary (Figure 6-11).

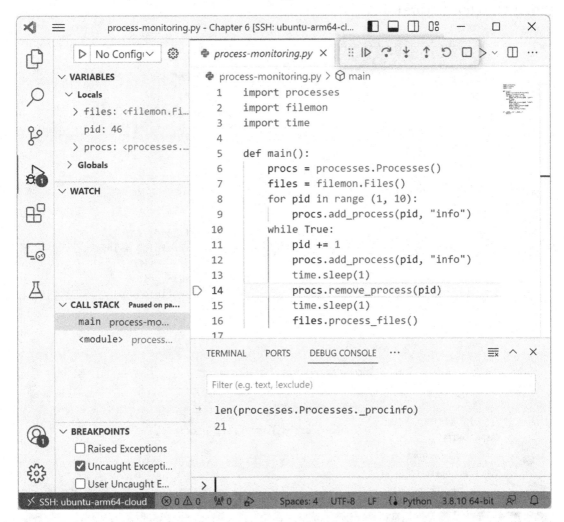

***Figure 6-11.*** *Using the debugging console to inspect variable values*

Since you know that the size of the dictionary starts increasing only after some time, open the processes.py file and put a **Code Breakpoint** on the first line of the add_ process function (Figure 6-12).

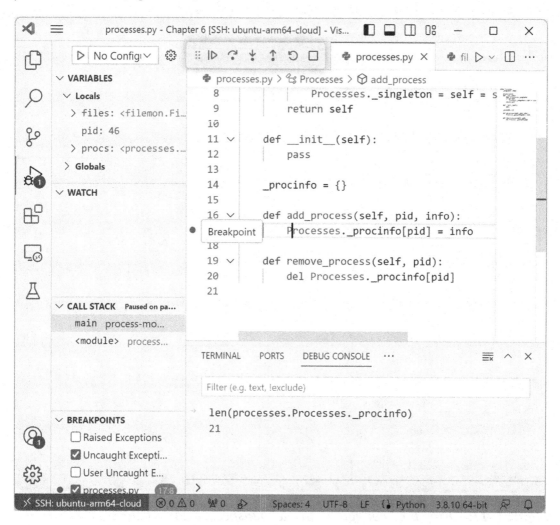

***Figure 6-12.***  *Setting a code breakpoint*

Then right-click the breakpoint dot and choose *Edit Breakpoint* and then specify the expression to be hit only when the dictionary size exceeds 100 (press Enter to set the condition). You will see the dot change in shape. Continue execution again using the *Debug* toolbar (F5) and wait (Figure 6-13).

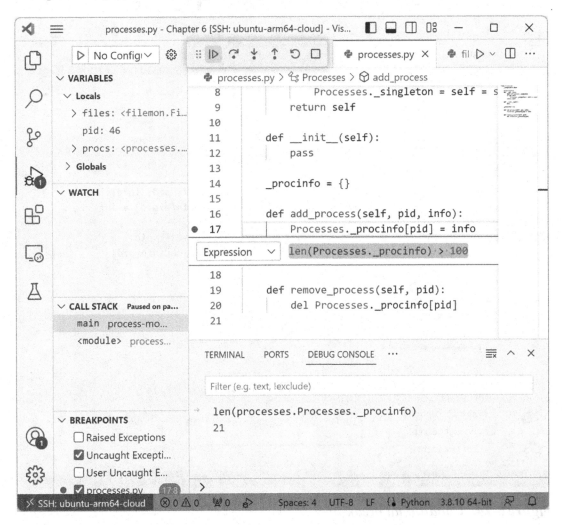

***Figure 6-13.*** *Setting the code breakpoint conditional expression*

After a few minutes, you had a new **Break-in** with the dot changed in shape. You can now repeat your **Variable Value** expression in the Debug Console. Please also note that the call stack has a reverse direction than the traceback (Figure 6-14).

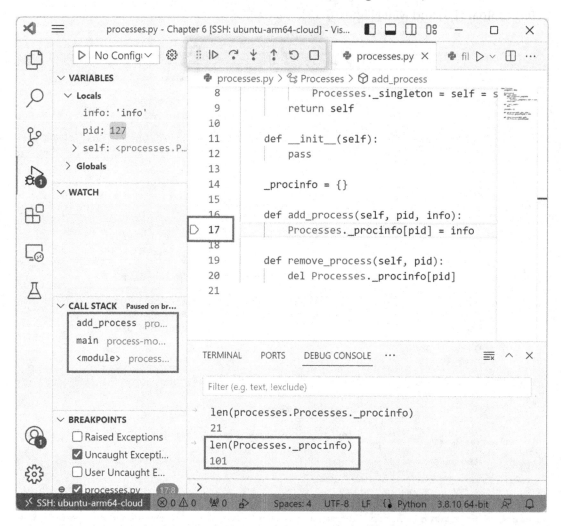

*Figure 6-14.* *Conditional break-in*

Now remove the conditional expression for the breakpoint and create another **Code Breakpoint** for the remove_process function with a **Breakpoint Action** in both breakpoints to log ADD and REMOVE messages (Figure 6-15).

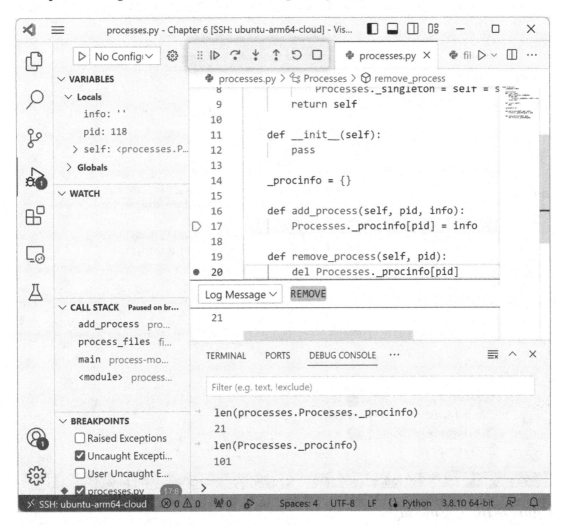

***Figure 6-15.***  *Specifying the breakpoint action*

Continue the execution and see that the log messages with ADD messages are twice as many as the REMOVE messages (Figure 6-16).

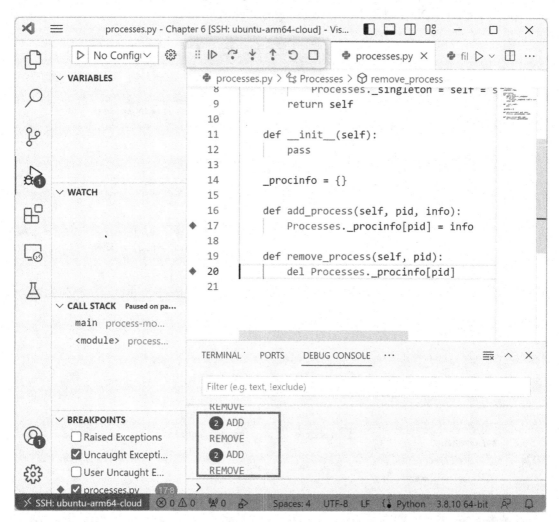

***Figure 6-16.*** *Log messages*

I don't know their **Usage Trace,** and I couldn't find a reliable way to print backtraces like I did when using PDB in the previous chapter. But since you are using the IDE editor, you can modify the file of interest directly and add additional debugging variables to print in the *Debugging Console* (Figure 6-17).

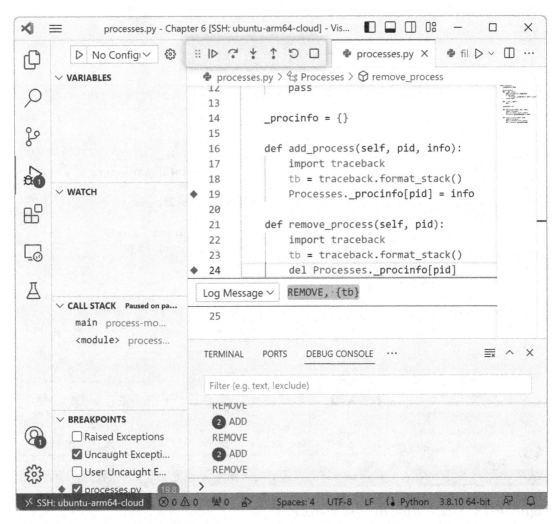

***Figure 6-17.*** *Adding traceback to logging*

You have to restart your debugging session from the *Debugging* toolbar (Ctrl-Shift-F5). After the restart, execution continues, and you wait for a few minutes, pause, and examine the last ADD and REMOVE entries in the log (Figure 6-18).

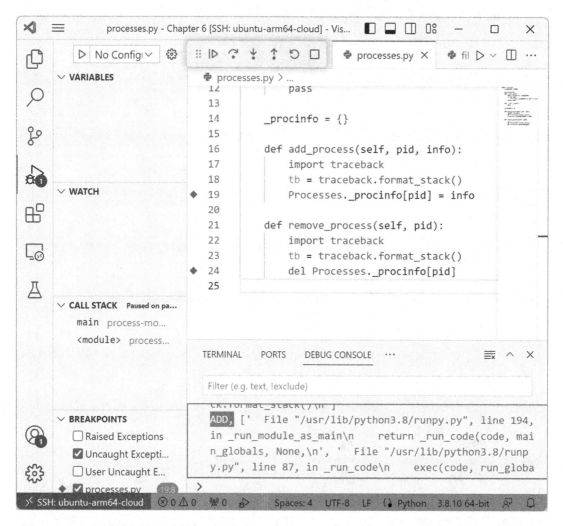

***Figure 6-18.*** *Generating the usage trace*

From the log, you can see that the ADD actions from the `filemon.py` module do not have any compensating REMOVE actions with the same recommendation as in the previous chapter.

...

**ADD,** [' File "/usr/lib/python3.8/runpy.py", line 194, in _run_module_as_ main\n    return _run_code(code, main_globals, None,\n', ' File "/usr/lib/

python3.8/runpy.py", line 87, in _run_code\n    exec(code, run_globals)\n',
' File "/home/ubuntu/.vscode-server/extensions/ms-python.python-2023.10.1/
pythonFiles/lib/python/debugpy/adapter/../../debugpy/launcher/../../
debugpy/__main__.py", line 39, in <module>\n    cli.main()\n', ' File
"/home/ubuntu/.vscode-server/extensions/ms-python.python-2023.10.1/
pythonFiles/lib/python/debugpy/adapter/../../debugpy/launcher/../../
debugpy/../debugpy/server/cli.py", line 430, in main\n    run()\n', ' File
"/home/ubuntu/.vscode-server/extensions/ms-python.python-2023.10.1/
pythonFiles/lib/python/debugpy/adapter/../../debugpy/launcher/../../
debugpy/../debugpy/server/cli.py", line 284, in run_file\n    runpy.
run_path(target, run_name="__main__")\n', ' File "/home/ubuntu/.vscode-
server/extensions/ms-python.python-2023.10.1/pythonFiles/lib/python/
debugpy/_vendored/pydevd/_pydevd_bundle/pydevd_runpy.py", line 321, in
run_path\n    return _run_module_code(code, init_globals, run_name,\n',
' File "/home/ubuntu/.vscode-server/extensions/ms-python.python-2023.10.1/
pythonFiles/lib/python/debugpy/_vendored/pydevd/_pydevd_bundle/pydevd_
runpy.py", line 135, in _run_module_code\n    _run_code(code, mod_globals,
init_globals,\n', ' File "/home/ubuntu/.vscode-server/extensions/ms-
python.python-2023.10.1/pythonFiles/lib/python/debugpy/_vendored/pydevd/_
pydevd_bundle/pydevd_runpy.py", line 124, in _run_code\n    exec(code,
run_globals)\n', ' File "/home/ubuntu/Python-Book/Chapter 6/process-
monitoring.py", line 19, in <module>\n    main()\n', ' File "/home/ubuntu/
Python-Book/Chapter 6/process-monitoring.py", line 16, in main\n    files.
process_files()\n', ' **File "/home/ubuntu/Python-Book/Chapter 6/filemon.
py"**, line 11, in process_files\n    self._processes.add_process(self._
count, "")\n', ' File "/home/ubuntu/Python-Book/Chapter 6/processes.py",
line 18, in add_process\n    tb = traceback.format_stack()\n']
**ADD**, [' File "/usr/lib/python3.8/runpy.py", line 194, in _run_module_as_
main\n    return _run_code(code, main_globals, None,\n', ' File "/usr/lib/
python3.8/runpy.py", line 87, in _run_code\n    exec(code, run_globals)\n',
' File "/home/ubuntu/.vscode-server/extensions/ms-python.python-2023.10.1/
pythonFiles/lib/python/debugpy/adapter/../../debugpy/launcher/../../
debugpy/__main__.py", line 39, in <module>\n    cli.main()\n', ' File
"/home/ubuntu/.vscode-server/extensions/ms-python.python-2023.10.1/
pythonFiles/lib/python/debugpy/adapter/../../debugpy/launcher/../../

debugpy/../debugpy/server/cli.py", line 430, in main\n    run()\n', ' File
"/home/ubuntu/.vscode-server/extensions/ms-python.python-2023.10.1/
pythonFiles/lib/python/debugpy/adapter/../../debugpy/launcher/../../
debugpy/../debugpy/server/cli.py", line 284, in run_file\n    runpy.
run_path(target, run_name="__main__")\n', ' File "/home/ubuntu/.vscode-
server/extensions/ms-python.python-2023.10.1/pythonFiles/lib/python/
debugpy/_vendored/pydevd/_pydevd_bundle/pydevd_runpy.py", line 321, in
run_path\n    return _run_module_code(code, init_globals, run_name,\n',
' File "/home/ubuntu/.vscode-server/extensions/ms-python.python-2023.10.1/
pythonFiles/lib/python/debugpy/_vendored/pydevd/_pydevd_bundle/pydevd_
runpy.py", line 135, in _run_module_code\n    _run_code(code, mod_globals,
init_globals,\n', ' File "/home/ubuntu/.vscode-server/extensions/ms-
python.python-2023.10.1/pythonFiles/lib/python/debugpy/_vendored/pydevd/_
pydevd_bundle/pydevd_runpy.py", line 124, in _run_code\n    exec(code,
run_globals)\n', ' File "/home/ubuntu/Python-Book/Chapter 6/process-
monitoring.py", line 19, in <module>\n    main()\n', ' **File "/home/ubuntu/
Python-Book/Chapter 6/process-monitoring.py"**, line 12, in main\n    procs.
add_process(pid, "info")\n', ' File "/home/ubuntu/Python-Book/Chapter
6/processes.py", line 18, in add_process\n    tb = traceback.format_
stack()\n']
**REMOVE**, [' File "/usr/lib/python3.8/runpy.py", line 194, in _run_module_
as_main\n    return _run_code(code, main_globals, None,\n', ' File "/
usr/lib/python3.8/runpy.py", line 87, in _run_code\n    exec(code, run_
globals)\n', ' File "/home/ubuntu/.vscode-server/extensions/ms-python.
python-2023.10.1/pythonFiles/lib/python/debugpy/adapter/../../debugpy/
launcher/../../debugpy/__main__.py", line 39, in <module>\n    cli.
main()\n', ' File "/home/ubuntu/.vscode-server/extensions/ms-python.
python-2023.10.1/pythonFiles/lib/python/debugpy/adapter/../../
debugpy/launcher/../../debugpy/../debugpy/server/cli.py", line 430, in
main\n    run()\n', ' File "/home/ubuntu/.vscode-server/extensions/ms-
python.python-2023.10.1/pythonFiles/lib/python/debugpy/adapter/../../
debugpy/launcher/../../debugpy/../debugpy/server/cli.py", line 284, in run_
file\n    runpy.run_path(target, run_name="__main__")\n', ' File "/home/
ubuntu/.vscode-server/extensions/ms-python.python-2023.10.1/pythonFiles/
lib/python/debugpy/_vendored/pydevd/_pydevd_bundle/pydevd_runpy.py",

line 321, in run_path\n    return _run_module_code(code, init_globals,
run_name,\n', '  File "/home/ubuntu/.vscode-server/extensions/ms-python.
python-2023.10.1/pythonFiles/lib/python/debugpy/_vendored/pydevd/_pydevd_
bundle/pydevd_runpy.py", line 135, in _run_module_code\n    _run_code(code,
mod_globals, init_globals,\n', '  File "/home/ubuntu/.vscode-server/
extensions/ms-python.python-2023.10.1/pythonFiles/lib/python/debugpy/_
vendored/pydevd/_pydevd_bundle/pydevd_runpy.py", line 124, in _run_
code\n    exec(code, run_globals)\n', '  File "/home/ubuntu/Python-Book/
Chapter 6/process-monitoring.py", line 19, in <module>\n    main()\n',
'  **File "/home/ubuntu/Python-Book/Chapter 6/process-monitoring.py"**, line
14, in main\n    procs.remove_process(pid)\n', '  File "/home/ubuntu/
Python-Book/Chapter 6/processes.py", line 23, in remove_process\n    tb =
traceback.format_stack()\n']
...

# Summary

In this chapter, you looked at debugging implementation patterns in the context of
the Python IDE and Visual Studio Code. You also learned how to connect it to WSL on
Windows and to a cloud environment using SSH. The next chapter looks at a few more
IDEs, including the one popular in machine learning environments; it also introduces
debugging presentation patterns.

# Debugging Presentation Patterns

In the previous chapter, you looked at debugging implementation patterns in the Python IDE Visual Studio Code context. In this chapter, you will look at another Python IDE, Jupyter Notebook and Lab, with a case study introducing a few more debugging analysis techniques and implementation patterns. Finally, after comparing Python CLIs and IDEs, you will discern a few basic debugging presentation patterns (Figure 7-1).

## Python Debugging Engines

Modern Python debugging IDEs have similar debugging engines; for example, Visual Studio Code uses debugpy[1], which can also be used in the CLI mode, similar to the pdb module used in Chapter 5. It also has remote debugging capabilities via listening on an interface and port, allowing debug clients to connect. PyCharm[2] uses pydevd[3] (which is also bundled with debugpy). Jupyter Notebook and Lab[4] use ipykernel[5] to access ipdb[6] debugging functionality provided by IPython[7].

---

[1] https://github.com/microsoft/debugpy
[2] https://www.jetbrains.com/pycharm/
[3] https://pypi.org/project/pydevd/
[4] https://jupyter.org/
[5] https://pypi.org/project/ipykernel/
[6] https://pypi.org/project/ipdb/
[7] https://ipython.org/

© Dmitry Vostokov 2024
D. Vostokov, *Python Debugging for AI, Machine Learning, and Cloud Computing*,
https://doi.org/10.1007/978-1-4842-9745-2_7

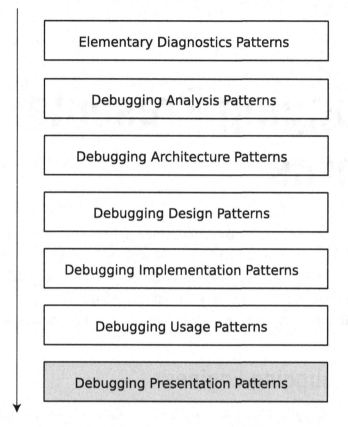

***Figure 7-1.*** *Pattern-oriented debugging process and debugging presentation patterns*

# Case Study

You will use a similar Python code for this case study as in the previous chapter but you will adapt it to the Jupyter Notebook format (Listing 7-1). Please note that Jupyter Notebook files can also be loaded and debugged in Visual Studio Code.

To recall the previous chapter, the case study models a problem I observed with a monitoring script that, after time, shows increasing process memory consumption.

***Listing 7-1.*** Memory Leak Example Adapted for Jupyter Notebook

```python
class Processes:
    _singleton = None

    @staticmethod
    def __new__(cls):
```

```python
        self = Processes._singleton
        if not self:
            Processes._singleton = self = super().__new__(cls)
        return self

    def __init__(self):
        pass

    _procinfo = {}

    def add_process(self, pid, info):
        Processes._procinfo[pid] = info

    def remove_process(self, pid):
        del Processes._procinfo[pid]

class Files:
    def __init__(self):
        self._processes = Processes()
        self._count = 0

    def process_files(self):
        self._count += 1
        if self._count > 25:
            self._processes.add_process(self._count, "")

import time

procs = Processes()
files = Files()
for pid in range (1, 10):
    procs.add_process(pid, "info")
while True:
    pid += 1
    procs.add_process(pid, "info")
    time.sleep(1)
    procs.remove_process(pid)
    time.sleep(1)
    files.process_files()
```

I suppose you are familiar with the Jupyter Notebook environment[8]. If not, you can install it with a simple command (I use Windows for this case study, Python 3.11.4, and assume your Python environment is accessible via PATH):

```
Chapter7>pip install notebook
...
```

To start the Jupyter Notebook environment, use the following command:

```
Chapter7>jupyter notebook
...
[I 2023-08-13 13:45:59.513 ServerApp] Jupyter Server 2.7.0 is running at:
[I 2023-08-13 13:45:59.513 ServerApp] http://localhost:8888/tree?token=2a
ae60bdf51574972f646f61f27a8bcb5416c08c2a835087
[I 2023-08-13 13:45:59.513 ServerApp]     http://127.0.0.1:8888/tree?token=
2aae60bdf51574972f646f61f27a8bcb5416c08c2a835087
[I 2023-08-13 13:45:59.514 ServerApp] Use Control-C to stop this server and
shut down all kernels (twice to skip confirmation).
[C 2023-08-13 13:45:59.675 ServerApp]

    To access the server, open this file in a browser:
...
    Or copy and paste one of these URLs:
        http://localhost:8888/tree?token=2aae60bdf51574972f646f61f27a8b
        cb5416c08c2a835087
...
```

On our system, a browser window opens automatically with a file tree window (Figure 7-2). Select the process-monitoring.ipynb file and open it. A new browser tab opens with the file's contents in a notebook cell (Figure 7-3). To execute any cell contents, use the toolbar or press <Shift-Enter> on the keyboard.

---

[8] Many developers use Anaconda or Miniconda environment (conda, https://docs.conda.io/projects/conda/en/stable/user-guide/install/download.html#anaconda-or-miniconda) and then use Jupyter Notebook there.

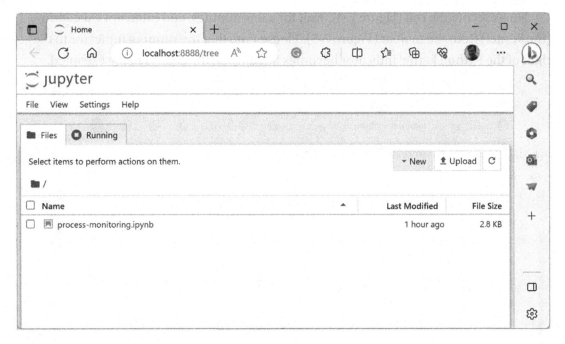

**Figure 7-2.** *Jupyter Notebook file tree window*

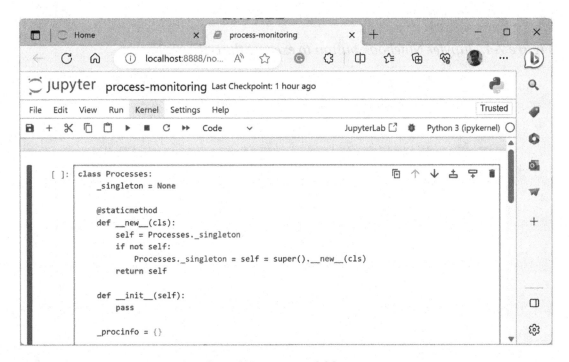

**Figure 7-3.** *Jupyter Notebook with an opened file*

Now execute the cell (Figure 7-4), wait for a few minutes, and interrupt execution (the kernel) from the toolbar (Figure 7-5). Please note that the running indicator to the right of the Python kernel indicator becomes filled when the code is being executed.

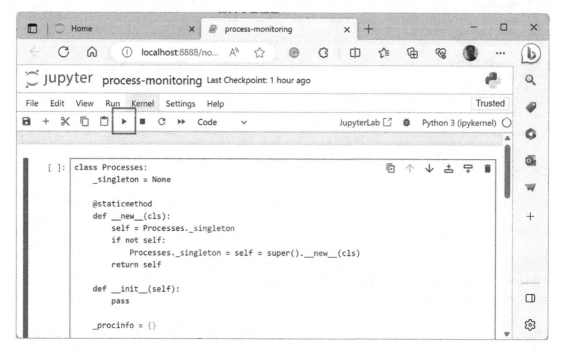

**Figure 7-4.** *Jupyter Notebook button to execute the cell*

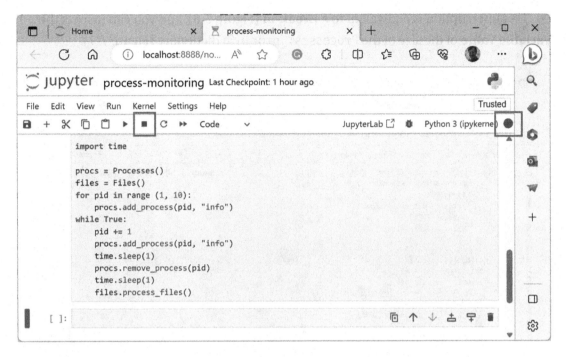

***Figure 7-5.*** *Jupyter Notebook button to stop cell execution*

This interruption is equivalent to a **Break-in** (Figure 7-6), and you can inspect the **Variable Value** of the size of the `Processes._procinfo` dictionary (Figure 7-7).

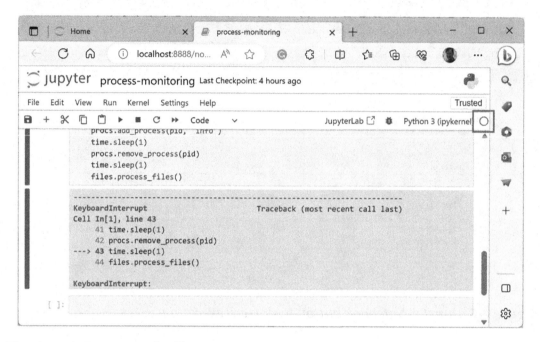

***Figure 7-6.*** *Interrupted cell*

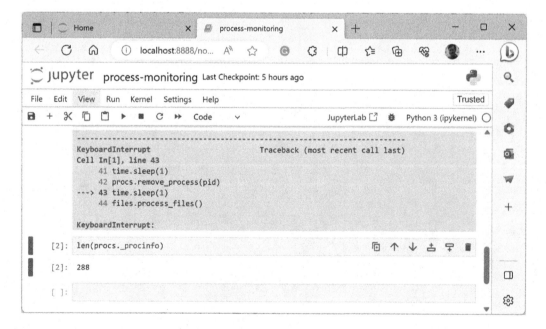

***Figure 7-7.*** *Inspecting variable values*

To put in **Code Breakpoints**, enable debugging (Figure 7-8). Enabling the debugger adds the gutter to cells where you can put breakpoints (Figure 7-9).

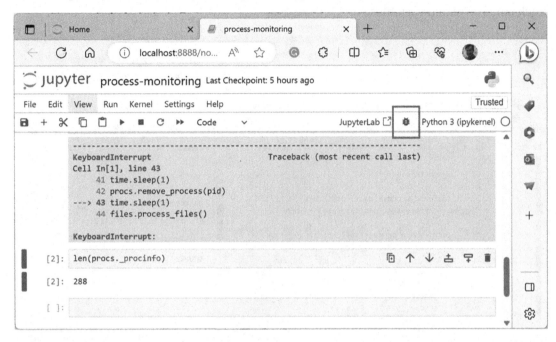

***Figure 7-8.*** *Enable the Debugger button*

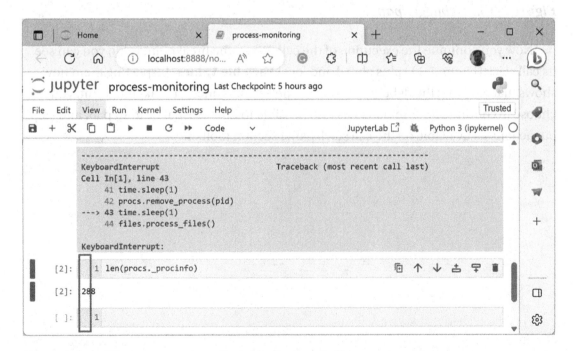

***Figure 7-9.*** *The gutter to turn breakpoints on/off*

You can now put two **Code Breakpoints** in the gutter and open the Debugger Panel via the View menu item (Figure 7-10), where you can see the list of breakpoints, the call stack, and the watch variables.

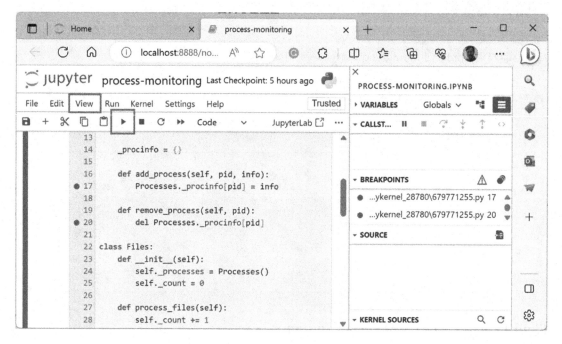

***Figure 7-10.*** *Debugger panel*

Now you continue the execution of the cell. When the breakpoint is hit, you can see its call stack (Figure 7-11) plus global and local **Variable Values**. Inspecting them now shows that enabling the debugger restarted the cell. But once a breakpoint is hit, you can choose to continue or terminate. It is also possible to do the usual next, step in, and step out actions all available on the call stack panel section toolbar.

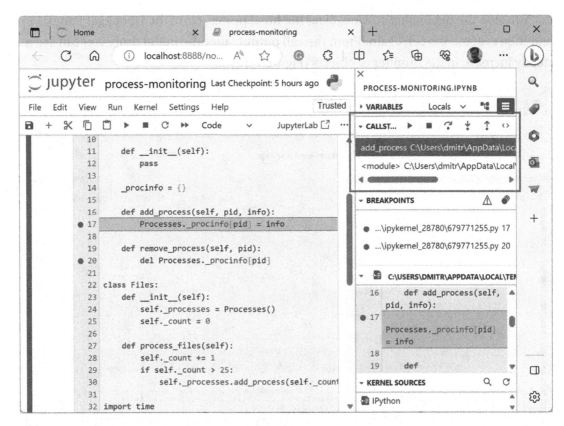

***Figure 7-11.*** *The breakpoint hit and call stack*

Let's now explore the garbage collection[9] (GC) stats and `tracemalloc` techniques[10] for **Memory Leak** debugging analysis and **Usage Trace** debugging implementation patterns.

For example, you can add the code from Listing 7-2 inside the `while` loop to check the number of allocated objects and other statistics.

***Listing 7-2.*** Code to Check the Number of Allocated Objects

```
import gc
print(len(gc.get_objects()))
print(gc.get_stats())
```

---

[9] Garbage collection (GC) manages memory automatically, ensuring that objects which are no longer needed are properly discarded to free up system resources. It helps when working with large data, such as when using Pandas.

[10] Brett Slatkin, *Effective Python*, Second Edition, p. 384 (ISBN-13: 978-0134853987)

You stop the debugging session, clear the breakpoints, close the Debugger Panel, and then execute the cell again. From the cell output, after the initial drop, you can see a steady increase in the number of objects (Figure 7-12).

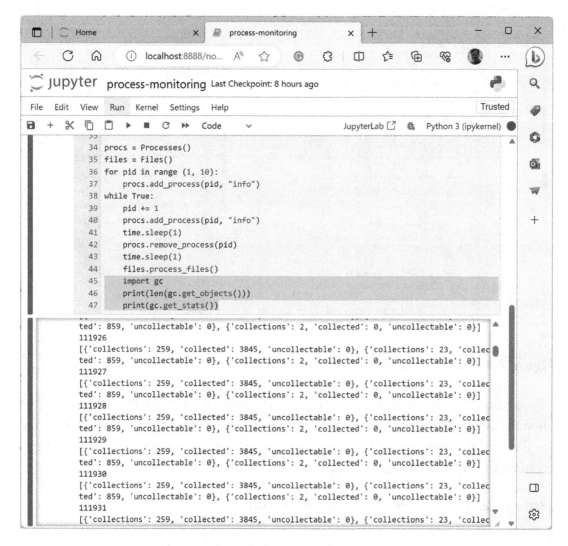

***Figure 7-12.*** *Tracing the number of objects*

The gc module has additional debugging flags that may help in other scenarios[11].

You can also see where all extra objects are allocated by using the tracemalloc module and comparing allocation snapshot differences. Replace the main script code with the code from Listing 7-3, where you also remove sleep calls for faster memory leak modeling.

***Listing 7-3.*** Code to Compare Allocations and Print the Topmost malloc User Traceback

```
procs = Processes()
files = Files()
import tracemalloc
tracemalloc.start()
s1 = tracemalloc.take_snapshot()
for pid in range (1, 10):
    procs.add_process(pid, "info")
try:
    while True:
        pid += 1
        procs.add_process(pid, "info")
        procs.remove_process(pid)
        files.process_files()
except:
    s2 = tracemalloc.take_snapshot()
    top_user = s2.compare_to(s1, "traceback")[0]
    print("\n".join(top_user.traceback.format()))
```

Run the cell again and interrupt it after a few seconds. Now, instead of the KeyboadInterrupt exception, you see the top allocation and its code (Figure 7-13).

---

[11] https://docs.python.org/3/library/gc.html

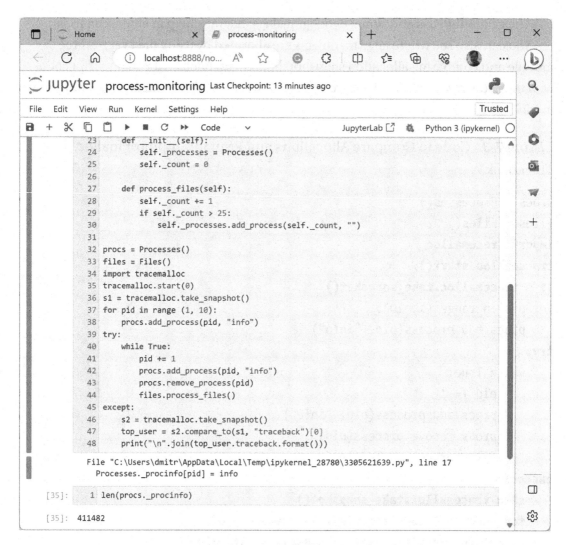

**Figure 7-13.** *The topmost allocation traceback*

You get the allocation distribution if you dump several top allocations instead of only the topmost. Using this technique, you can compare normal runs with those leaking process memory, the so-called **Object Distribution Anomaly** debugging analysis pattern.

Please check its extensive documentation for additional examples of the tracemalloc module usage[12].

_____

[12] https://docs.python.org/3/library/tracemalloc.html

# Suggested Presentation Patterns

So far, you have seen both command line and visual debuggers. I suggest the following presentation patterns:

- REPL: Read, execute, print loop

- Breakpoint Toolbar

- Action Toolbar

- State Dashboard

The meaning of these pattern names should be obvious. The **State Dashboard** includes panel windows for variables and stack traces. The **REPL** abbreviation traditionally meant **read, eval, and print loop**, but I changed eval to **execute** for debugging commands. The **Action Toolbar** is for actions intrinsic to debugging sessions, such as stepping. The **Breakpoint Toolbar** is for setting and removing breakpoints and specifying their actions.

# Summary

In this chapter, you looked at debugging engines for various IDEs, a Jupyter Notebook case study, and the most common debugging presentation patterns. The next two chapters introduce debugging architecture and debugging design patterns.

# CHAPTER 8

# Debugging Architecture Patterns

In the previous two chapters, you looked at some IDEs and debugging presentation patterns. You now have enough background material that I can introduce debugging architecture patterns (Figure 8-1) in this chapter. These are solutions to common problems of how to proceed with debugging. These patterns answer the most important questions of where, when, and how we debug and what we debug based on the debugging analysis patterns covered in Chapter 4. The diagnostics and debugging "questions" pattern language thinking originated from memory and trace acquisition patterns.[1] There are many ways to answer such questions and propose debugging strategies and methods; sometimes, you may even need to combine different approaches. However, there are the most important decisions[2] to make before any debugging is done, and we name them *debugging architecture patterns* vs. *debugging design patterns*, which can be many for the particular architecture pattern. All debugging architecture patterns can be combined to form pattern sequences and pattern networks to solve complex problems of software behavior.

---

[1] Dmitry Vostokov, *Theoretical Software Diagnostics: Collected Articles*, Third Edition, 2020 (ISBN-13: 978-1912636334), pp. 176-177

[2] Frank Buschmann et al., *Pattern-Oriented Software Architecture: On Patterns and Pattern Languages*, 2007 (ISBN-13: 978-0471486480)

© Dmitry Vostokov 2024
D. Vostokov, *Python Debugging for AI, Machine Learning, and Cloud Computing*,
https://doi.org/10.1007/978-1-4842-9745-2_8

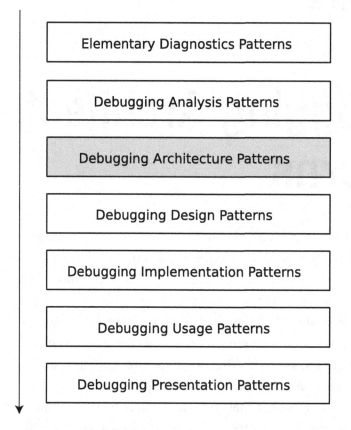

*Figure 8-1.*  *Pattern-oriented debugging process and debugging architecture patterns*

# The Where? Category

This category of debugging architecture patterns answers the question of **where** debugging session(s) will happen relative to the place of software execution. This is especially important because of the distributed nature of cloud-native applications and possible access restrictions.

# In Papyro

Some debugging scenarios may only need a whiteboard (or blackboard) to solve a problem, and some must be solved without access to production environments due to security restrictions. This pattern name is borrowed from experimental studies[3] where only paper and pencil are used for analysis. However, in our modern times, this also includes descriptions, screenshots, and various tools that may be used to view and search collected data, such as log viewers.

# In Vivo

This (micro)biological[4] pattern's name is borrowed from the proposed pattern language for POCs (Proof of Concept)[5]. It means that you debug in the environment where you have the original problem, such as the specific running process from the specific Python virtual directory.

# In Vitro

This pattern's name is borrowed from (micro)biology[6] and, in the debugging architecture context, it means that you try to reproduce and debug the problem in the environment that is outside of its original place, such as on a developer machine.

# In Silico

This pattern's name is borrowed from experimental sciences[7] and in this context means that you try to reproduce and debug the problem using some modeling process or system. For example, if particular library doesn't work in a production environment, you create a test environment with similar characteristics and input data where you run a small program that uses the library and see if the toy software model throws the same errors.

---

[3] https://en.wikipedia.org/wiki/In_papyro

[4] https://en.wikipedia.org/wiki/In_vivo

[5] Dmitry Vostokov, *Proof of Concept Engineering Patterns, Memory Dump Analysis Anthology*, Volume 15, 2023 (ISBN-13: 978-1912636150), pp. 284 - 285

[6] https://en.wikipedia.org/wiki/In_vitro

[7] https://en.wikipedia.org/wiki/In_silico

# In Situ

Situ patterns[8] address the question where you analyze the execution artifacts, such as logs. In the case of **In Situ**, you keep artifacts where they were produced and analyze them right there. This pattern also covers examples of debugging using console logs from distributed environments such as Kubernetes.

# Ex Situ

**Ex Situ** is the opposite of **In Situ:** you move produced artifacts out of the problem environment for offline analysis on a different computer. In cloud environments, observability logs are usually transported to some log collection services for later analysis if necessary. This pattern also covers crash and hang memory dumps or heap snapshots gathered for memory leak analysis later in a different environment where all necessary visual tools are installed.

# The When? Category

This category of debugging architecture patterns answers the question of **when** debugging session(s) should and will happen relative to the timeline of software execution.

# Live

**Live** debugging usually starts before the problem manifests itself. You either start the program under a debugger or attach a debugger at some point when you think the problem will appear soon or after some action. Then you wait for the problem to happen while periodically tracing execution progress and inspecting program state such as variables.

---

[8] https://en.wikipedia.org/wiki/In_situ

# JIT

Just-in-time (**JIT**) debugging starts at the moment the problem manifests itself. For example, the runtime environment or operating system launches the debugger as a response to some signal or exception.

## Postmortem

**Postmortem** debugging starts after the problem manifests itself, such as after a crash or when various execution artifacts, such as traces and logs, are produced to be debugged either **In Situ** or **Ex Situ**. The debugging is then done usually **In Papyro** mode, assisted by software tools such as debuggers for viewing memory dumps and log viewers.

Time travel debugging[9] usually blends **Postmortem** and **Live** as some tools on specific operating system platforms may allow recording program execution and replaying it later by going back in time to do test state corrections and different execution paths.

# The What? Category

This category of debugging architecture patterns answers the question of **what** you will debug during the debugging session(s) relative to the structure of the program.

## Code

Fixing **code** defects is the traditional goal of debugging activity. Here you debug code in the traditional sense by tracing source lines, making changes to the state, and altering execution paths. This is usually done by using source code debuggers and Python language and ecosystem tracing and logging facilities.

---

[9] https://en.wikipedia.org/wiki/Time_travel_debugging

# Data

There can also be problems with configuration or input data, and the goal of data debugging is to fix such **Data** defects. This is also important for machine learning using already well-debugged libraries and frameworks.

# Interaction

Sometimes, the problem is not in code or data but in the improper usage of the software. Here **Interaction** defects may be analyzed by HCI action recorders or simply screen recording.

# The How? Category

This category of debugging architecture patterns answers the question of **how** you will debug relative to software execution time and space.

## Software Narrative

**Software Narratives** are software execution artifacts not limited to traces and logs. They also include source code and various documents and descriptions. This term was originally introduced in software narratology[10]. You choose this architectural pattern if you want to see how the software structure and behavior changes with time but you are not interested in seeing all state changes in every part of the program.

## Software State

The traditional live and postmortem debugging approaches inspect the program's state, such as memory. For example, they can watch the value of a variable or object field over time (live debugging) or inspect memory addresses at the time of a crash or hang (postmortem debugging). You use this **Software State** architectural pattern when you want to analyze the program or even the entire system state at once. A typical example is

---

[10] Dmitry Vostokov, *Software Narratology: A Definition, Theoretical Software Diagnostics: Collected Articles*, Third Edition, 2020 (ISBN-13: 978-1912636334), p. 58

a program that depends on several system services, and these services depend on each other and the operating system. In this case, you may need the whole physical memory snapshot for the analysis of different virtual memory process spaces.

# Summary

In this chapter, you overviewed the debugging architecture patterns. You will encounter them again in some of the case studies in subsequent chapters. The next chapter looks at the most common debugging design patterns that are used to refine more general debugging architecture patterns in the context of case studies that involve the analysis, architecture, and design of debugging.

# CHAPTER 9

# Debugging Design Patterns

In the previous chapter, you looked at debugging architecture patterns, answering the most important question of where, when, and how we debug and what we debug.

In this chapter, I introduce the most common debugging design patterns (Figure 9-1) in real-life case studies involving elementary diagnostics patterns, analysis, and debugging architecture. These patterns are **Mutation**, **Replacement**, **Isolation**, and **Try-Catch**. The selection of pattern names is based on the author's experience with debugging in software domains, including system programming, user interface development, machine learning, and cloud-native applications. Other software domains, such as games or embedded, may have different and additional pattern names. Pattern languages are not fixed; everyone can have their patterns or adapt existing patterns from the book you are reading if this helps their teams and software community[1].

**Mutation** deals with changing code, data, and configuration values to see the outcome during execution and while using debugging implementation patterns.

**Replacement** works on a higher level where you can replace whole functions, classes, or even components with others that have compatible interfaces.

**Isolation** is used to debug code independently from other parts while having all other dependencies fixed and not having any side effects on the code you debug.

**Try-Catch** is a procedure where you repeat all chosen patterns and hope to see changes in behavior.

---

[1] Robert Nystrom, "Introduction," *Game Programming Patterns*, 2014 (ISBN-13: 978-0990582908)

© Dmitry Vostokov 2024
D. Vostokov, *Python Debugging for AI, Machine Learning, and Cloud Computing*,
https://doi.org/10.1007/978-1-4842-9745-2_9

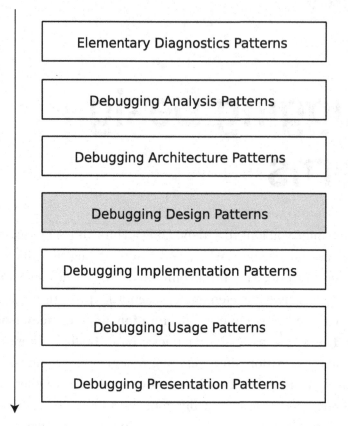

***Figure 9-1.*** *Pattern-oriented debugging process and debugging design patterns*

You may have had the impression that the pattern-oriented debugging process is straightforward and waterfall-like. This impression is far from the real case debugging scenarios of complex software issues. The actual pattern-oriented debugging process is usually iterative. It may trigger the reevaluation of debugging requirements, diagnostics (analysis), architecture, and design based on debugging results and a new set of execution artifacts (Figure 9-2) with new sets of corresponding debugging patterns for use.

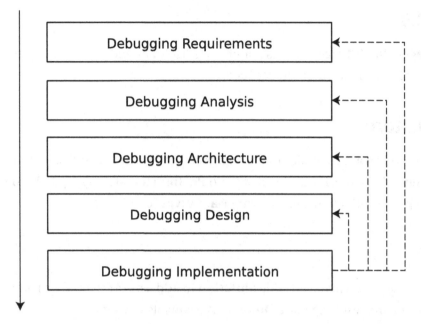

*Figure 9-2.* *Iterative nature of the pattern-oriented debugging process*

Let's now examine two debugging case studies and their corresponding debugging pattern layers.

# CI Build Case Study

This case study describes the debugging problem one of the engineers faced when a cloud CI build system written in Python failed after the upgrade. The problem had to be fixed quickly for the time being to do builds now without escalating code change to the cloud CI build system team and waiting for a new release.

## Elementary Diagnostics

The system crashed, printing a stack trace on the console with the following bottom line:

```
...
  File ".../ci/build-driver.py", line 346, in isVersion
    varStr = variable.replace("-","")
AttributeError: 'Variable' object has no attribute 'replace'
```

## Analysis

The **Managed Stack Trace** analysis pattern was useful in finding the problem object variable with the failing method replace.

## Architecture

The Python code was unfamiliar to the engineer and executed inside a Python virtual environment. Therefore, it was decided to do **Postmortem** debugging **In Vivo** inside the same failing environment using **Software Narratives** (console logs).

## Design

The debugging design involved code **Mutation** to add a print(variable) statement and then analyzing the resulting **State Dump** in the console output.

## Implementation

The engineer used the **Type Structure** and the **Variable Value** to find out the best course of action. After the code change, the console output included the representation of the variable object, which contained the name of configuration file parameter instead of value. The build system configuration file was written in one of the popular cloud-native configuration languages. So the engineer changed the debugging design, this time with the configuration **Mutation,** by replacing all occurrences of the parameter variable in the build configuration file with a hard-coded value as a temporary fix. This time the CI build was successful.

# Data Processing Case Study

This case study describes the debugging problem the engineers faced when using the Pandas data analysis and manipulation library in a third-party data processing and machine learning system on Windows that used Python DLL to execute the Python scripts.

# Elementary Diagnostics

In addition to the ability to execute Python scripts, the system also had embedded Python REPL. However, REPL experienced a **Hang** when executing the following Python code line:

```
import pandas
```

Since the data processing system was frozen, it was decided to save its process memory dump as described in Chapter 3's "How to Generate Process Memory Dumps on Windows" section.

# Analysis

To analyze the memory dump file artifact and its **Stack Trace Collection** and then find the problem **Stack Trace** and, ultimately, the **Origin Module**, engineers used the Microsoft debugger, WinDbg.

If you have never used WinDbg but are doing development on Windows, it is time to install this debugging tool. I recommend the WinDbg app (former WinDbg Preview), which can be downloaded and installed from the Microsoft site[2].

The collected process memory dump showed the following problem stack trace (the ~* command means to execute the next kcL command for each thread; the kcL command means print stack trace in a short form and without source code references for visual clarity here):

```
0:000> ~*kcL
  11  Id: 5bb0.7818 Suspend: 1 Teb: 000000dc`e1f28000 Unfrozen
 # Call Site
00 ntdll!NtWaitForSingleObject
01 KERNELBASE!WaitForSingleObjectEx
02 python37!_PyCOND_WAIT_MS
03 python37!PyCOND_TIMEDWAIT
04 python37!take_gil
05 python37!PyEval_RestoreThread
06 python37!PyGILState_Ensure
07 _multiarray_umath_cp37_win_amd64!PyInit__multiarray_umath
```

---

[2] https://learn.microsoft.com/en-gb/windows-hardware/drivers/debugger/

```
08 _multiarray_umath_cp37_win_amd64!PyInit__multiarray_umath
09 _multiarray_umath_cp37_win_amd64
0a _multiarray_umath_cp37_win_amd64
0b _multiarray_umath_cp37_win_amd64
0c _multiarray_umath_cp37_win_amd64
0d _multiarray_umath_cp37_win_amd64!PyInit__multiarray_umath
0e python37!type_call
0f python37!_PyObject_FastCallKeywords
10 python37!call_function
11 python37!_PyEval_EvalFrameDefault
12 python37!PyEval_EvalFrameEx
13 python37!function_code_fastcall
14 python37!_PyFunction_FastCallKeywords
15 python37!call_function
16 python37!_PyEval_EvalFrameDefault
17 python37!PyEval_EvalFrameEx
18 python37!_PyEval_EvalCodeWithName
19 python37!PyEval_EvalCodeEx
1a python37!PyEval_EvalCode
1b python37!builtin_exec_impl
1c python37!builtin_exec
1d python37!_PyMethodDef_RawFastCallDict
1e python37!_PyEval_EvalFrameDefault
1f python37!PyEval_EvalFrameEx
20 python37!_PyEval_EvalCodeWithName
21 python37!_PyFunction_FastCallKeywords
22 python37!call_function
23 python37!_PyEval_EvalFrameDefault
24 python37!PyEval_EvalFrameEx
25 python37!function_code_fastcall
26 python37!_PyFunction_FastCallKeywords
27 python37!call_function
28 python37!_PyEval_EvalFrameDefault
29 python37!PyEval_EvalFrameEx
2a python37!function_code_fastcall
```

```
2b python37!_PyFunction_FastCallKeywords
2c python37!call_function
2d python37!_PyEval_EvalFrameDefault
2e python37!PyEval_EvalFrameEx
2f python37!function_code_fastcall
30 python37!_PyFunction_FastCallKeywords
31 python37!call_function
32 python37!_PyEval_EvalFrameDefault
33 python37!PyEval_EvalFrameEx
34 python37!function_code_fastcall
35 python37!_PyFunction_FastCallDict
36 python37!_PyObject_FastCallDict
37 python37!object_vacall
38 python37!_PyObject_CallMethodIdObjArgs
39 python37!import_find_and_load
3a python37!PyImport_ImportModuleLevelObject
3b python37!_PyEval_EvalFrameDefault
3c python37!PyEval_EvalFrameEx
3d python37!_PyEval_EvalCodeWithName
3e python37!PyEval_EvalCodeEx
3f python37!PyEval_EvalCode
40 python37!builtin_exec_impl
41 python37!builtin_exec
42 python37!_PyMethodDef_RawFastCallDict
43 python37!_PyEval_EvalFrameDefault
44 python37!PyEval_EvalFrameEx
45 python37!_PyEval_EvalCodeWithName
46 python37!_PyFunction_FastCallKeywords
47 python37!call_function
48 python37!_PyEval_EvalFrameDefault
49 python37!PyEval_EvalFrameEx
4a python37!function_code_fastcall
4b python37!_PyFunction_FastCallKeywords
4c python37!call_function
4d python37!_PyEval_EvalFrameDefault
```

```
4e python37!PyEval_EvalFrameEx
4f python37!function_code_fastcall
50 python37!_PyFunction_FastCallKeywords
51 python37!call_function
52 python37!_PyEval_EvalFrameDefault
53 python37!PyEval_EvalFrameEx
54 python37!function_code_fastcall
55 python37!_PyFunction_FastCallKeywords
56 python37!call_function
57 python37!_PyEval_EvalFrameDefault
58 python37!PyEval_EvalFrameEx
59 python37!function_code_fastcall
5a python37!_PyFunction_FastCallDict
5b python37!_PyObject_FastCallDict
5c python37!object_vacall
5d python37!_PyObject_CallMethodIdObjArgs
5e python37!import_find_and_load
5f python37!PyImport_ImportModuleLevelObject
60 python37!_PyEval_EvalFrameDefault
61 python37!PyEval_EvalFrameEx
62 python37!_PyEval_EvalCodeWithName
63 python37!PyEval_EvalCodeEx
64 python37!PyEval_EvalCode
65 python37!builtin_exec_impl
66 python37!builtin_exec
67 python37!_PyMethodDef_RawFastCallDict
68 python37!_PyEval_EvalFrameDefault
69 python37!PyEval_EvalFrameEx
6a python37!_PyEval_EvalCodeWithName
6b python37!_PyFunction_FastCallKeywords
6c python37!call_function
6d python37!_PyEval_EvalFrameDefault
6e python37!PyEval_EvalFrameEx
6f python37!function_code_fastcall
70 python37!_PyFunction_FastCallKeywords
```

```
71  python37!call_function
72  python37!_PyEval_EvalFrameDefault
73  python37!PyEval_EvalFrameEx
74  python37!function_code_fastcall
75  python37!_PyFunction_FastCallKeywords
76  python37!call_function
77  python37!_PyEval_EvalFrameDefault
78  python37!PyEval_EvalFrameEx
79  python37!function_code_fastcall
7a  python37!_PyFunction_FastCallKeywords
7b  python37!call_function
7c  python37!_PyEval_EvalFrameDefault
7d  python37!PyEval_EvalFrameEx
7e  python37!function_code_fastcall
7f  python37!_PyFunction_FastCallDict
80  python37!_PyObject_FastCallDict
81  python37!object_vacall
82  python37!_PyObject_CallMethodIdObjArgs
83  python37!import_find_and_load
84  python37!PyImport_ImportModuleLevelObject
85  python37!_PyEval_EvalFrameDefault
86  python37!PyEval_EvalFrameEx
87  python37!_PyEval_EvalCodeWithName
88  python37!PyEval_EvalCodeEx
89  python37!PyEval_EvalCode
8a  python37!builtin_exec_impl
8b  python37!builtin_exec
8c  python37!_PyMethodDef_RawFastCallDict
8d  python37!_PyEval_EvalFrameDefault
8e  python37!PyEval_EvalFrameEx
8f  python37!_PyEval_EvalCodeWithName
90  python37!_PyFunction_FastCallKeywords
91  python37!call_function
92  python37!_PyEval_EvalFrameDefault
93  python37!PyEval_EvalFrameEx
```

```
94 python37!function_code_fastcall
95 python37!_PyFunction_FastCallKeywords
96 python37!call_function
97 python37!_PyEval_EvalFrameDefault
98 python37!PyEval_EvalFrameEx
99 python37!function_code_fastcall
9a python37!_PyFunction_FastCallKeywords
9b python37!call_function
9c python37!_PyEval_EvalFrameDefault
9d python37!PyEval_EvalFrameEx
9e python37!function_code_fastcall
9f python37!_PyFunction_FastCallKeywords
a0 python37!call_function
a1 python37!_PyEval_EvalFrameDefault
a2 python37!PyEval_EvalFrameEx
a3 python37!function_code_fastcall
a4 python37!_PyFunction_FastCallDict
a5 python37!_PyObject_FastCallDict
a6 python37!object_vacall
a7 python37!_PyObject_CallMethodIdObjArgs
a8 python37!import_find_and_load
a9 python37!PyImport_ImportModuleLevelObject
aa python37!_PyEval_EvalFrameDefault
ab python37!PyEval_EvalFrameEx
ac python37!_PyEval_EvalCodeWithName
ad python37!PyEval_EvalCodeEx
ae python37!PyEval_EvalCode
af python37!builtin_exec_impl
b0 python37!builtin_exec
b1 python37!_PyMethodDef_RawFastCallKeywords
b2 python37!_PyCFunction_FastCallKeywords
b3 python37!call_function
b4 python37!_PyEval_EvalFrameDefault
b5 python37!PyEval_EvalFrameEx
```

```
b6 python37!function_code_fastcall
b7 python37!_PyFunction_FastCallKeywords
b8 python37!call_function
b9 python37!_PyEval_EvalFrameDefault
ba python37!PyEval_EvalFrameEx
bb python37!_PyEval_EvalCodeWithName
bc python37!_PyFunction_FastCallKeywords
bd python37!call_function
be python37!_PyEval_EvalFrameDefault
bf python37!PyEval_EvalFrameEx
c0 python37!function_code_fastcall
c1 python37!_PyFunction_FastCallKeywords
c2 python37!call_function
c3 python37!_PyEval_EvalFrameDefault
c4 python37!PyEval_EvalFrameEx
c5 python37!_PyEval_EvalCodeWithName
c6 python37!_PyFunction_FastCallKeywords
c7 python37!call_function
c8 python37!_PyEval_EvalFrameDefault
c9 python37!PyEval_EvalFrameEx
ca python37!_PyEval_EvalCodeWithName
cb python37!PyEval_EvalCodeEx
cc python37!PyEval_EvalCode
cd python37!run_mod
ce python37!PyRun_StringFlags
cf python37!PyRun_String
d0 data_module!execute_script
...
dc KERNEL32!BaseThreadInitThunk
dd ntdll!RtlUserThreadStart
...
```

It looked like the thread using the _multiarray_umath_cp37_win_amd64 DLL was blocked, waiting for the Global Interpreter Lock (GIL). The following information was available about that module:

```
0:000> lmv m _multiarray_umath_cp37_win_amd64
Browse full module list
start             end                 module name
00007ff8`e6f40000 00007ff8`e722b000   _multiarray_umath_cp37_win_amd64
C (export symbols)        C:\Python37\lib\site-packages\numpy\core\_
multiarray_umath.cp37-win_amd64.pyd
    Loaded symbol image file: C:\Python37\lib\site-packages\numpy\core\_
    multiarray_umath.cp37-win_amd64.pyd
    Image path: C:\Python37\lib\site-packages\numpy\core\_multiarray_umath.
    cp37-win_amd64.pyd
    Image name: _multiarray_umath.cp37-win_amd64.pyd
    Browse all global symbols  functions  data
    Timestamp:        Tue Apr 12 03:42:29 2022 (6254E715)
    CheckSum:         00000000
    ImageSize:        002EB000
    Translations:     0000.04b0 0000.04e4 0409.04b0 0409.04e4
    Information from resource tables:
```

It looked like this DLL came from the NumPy Python library. It was also found that the issue looked like a known issue when importing NumPy[3]. This problem was also confirmed by importing other packages with the NumPy dependency; the program also became frozen.

# Architecture

To speed up the debugging process and avoid frequently sending memory dumps, engineers proposed using **In Vivo JIT Code** debugging by attaching the WinDbg debugger[4] to the process once the program became frozen and then analyzing its **Software State**.

---

[3] https://mail.python.org/pipermail/python-dev/2019-January/156096.html

[4] Sometimes the developer needs to be careful not to inadvertently corrupt the program state of the running program when attaching and detaching the debugger. The WinDbg debugger has a

## Design

Since the problem involved the Python interpreter and native libraries from the third-party package, it was proposed to Python version **Replacement**. If no compatible Python helps (for example, the system couldn't use the latest Python due to other legacy components), then you can try a **Replacement** for the Pandas library. Find a compatible library with a similar interface that doesn't use NumPy.

## Implementation

Debugging implementation patterns involved a **Break-in** and **Breakpoint Action** to print stack traces. Finally, after numerous attempts with different Python versions and libraries with similar interfaces, it was decided to use the Polars[5] library instead of Pandas.

## Summary

In this chapter, you explored some case studies and basic debugging patterns. The next chapter examines pragmatics: debugging usage patterns, the last debugging pattern category.

---

non-invasive attach option.

[5] https://pypi.org/project/polars/

# CHAPTER 10

# Debugging Usage Patterns

In the previous chapter, you looked at debugging design patterns. In this chapter, you will explore the most common debugging usage patterns (Figure 10-1).

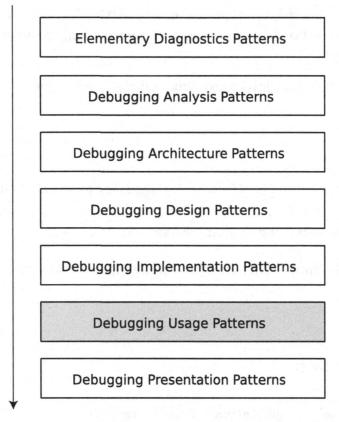

**Figure 10-1.** *Pattern-oriented debugging process and debugging design patterns*

© Dmitry Vostokov 2024
D. Vostokov, *Python Debugging for AI, Machine Learning, and Cloud Computing,*
https://doi.org/10.1007/978-1-4842-9745-2_10

These usage patterns are about debugging pragmatics and automating patterns from other layers. They accumulate experience developing and using debugging tools and products across various operating systems and platforms. I list them with brief descriptions below.

# Exact Sequence

Sometimes pattern implementations for the specific debugger, operating system, or debugging space involve exact sequences of commands or specific command options to achieve the desired effect. Not following **Exact Sequences** may generate unintended side effects and incorrect and misleading command output.

Typical examples are

- Switching to the correct process space and reloading user space symbols in live Windows or complete memory dump postmortem debugging

- Dumping symbolic references from a thread stack region

# Scripting

Most debuggers include support for scripting languages; for example, GDB provides support for simple scripting (Listings 10-1 and 10-2) and Python scripts[1], and WinDbg includes native scripting language[2] and also supports JavaScript[3].

***Listing 10-1.*** A Simple GDB Script for Double Memory Dereference

```
define dpp
    set $i = 0
    set $p = $arg0
    while $i < $arg1
```

---

[1] https://sourceware.org/gdb/onlinedocs/gdb/Python.html

[2] Introduction to WinDbg Scripts for C/C++ Users, www.dumpanalysis.org/WCDA/WCDA-Sample-Chapter.pdf

[3] Dmitry Vostokov, *Advanced Windows Memory Dump Analysis with Data Structures*, Fourth Edition, Revised, 2022 (ISBN-13: 978-1912636778)

```
        printf "%p: ", $p
        x/ga *(long *)$p
        set $i = $i + 1
        set $p = $p + 8
    end
end
```

*Listing 10-2.* A Sample Output from the GDB Script Example

(gdb) **source script.txt**

(gdb) **dpp 0x7ffdf45637f8 10**

```
0x7ffdf45637f8: 0x7ffdf4565756: 0x622f3d4c4c454853
0x7ffdf4563800: 0x7ffdf4565766: 0x544e4f4354534948
0x7ffdf4563808: 0x7ffdf456577d: 0x545349445f4c5357
0x7ffdf4563810: 0x7ffdf4565794: 0x5345443d454d414e
0x7ffdf4563818: 0x7ffdf45657a9: 0x6d6f682f3d445750
0x7ffdf4563820: 0x7ffdf45657c7: 0x3d454d414e474f4c
0x7ffdf4563828: 0x7ffdf45657d8: 0x4944504d545f434d
0x7ffdf4563830: 0x7ffdf45657f3: 0x313d4449535f434d
0x7ffdf4563838: 0x7ffdf45657fe: 0x6f682f3d454d4f48
0x7ffdf4563840: 0x7ffdf4565812: 0x5f6e653d474e414c
```

# Debugger Extension

Usually, by a debugger extension, we mean a DLL or shared library that a debugger process can load to provide additional debugging commands or extend the existing ones. GDB uses Python scripting capabilities for the Python debugging extension[4], which you used in the Chapter 4 case study. WinDbg has many extensions[5] and provides

---

[4] https://devguide.python.org/development-tools/gdb/index.html

[5] https://github.com/anhkgg/awesome-windbg-extensions

C/C++ APIs to write them[6], but the dedicated third-party extension to debug Python doesn't work with the latest Python versions[7]. You will use WinDbg extensions for native debugging analysis patterns in the next chapter.

# Abstract Command

Sometimes, when describing or communicating debugging scenarios, it is easy to abstract differences between debuggers, their commands, and **Exact Sequences** by creating a debugging domain language and using either debugging extensions or scripts for this language implementation. This approach is adopted in some memory forensics tools.

# Space Translation

Debugging implementation pattern names used for debugging process user space can be reused for kernel or managed spaces, although their implementations, such as debugger commands, may differ. A typical example is the need to dump symbolic references from user space thread stack and kernel space thread stack memory regions.

# Lifting

Problem solving techniques can be borrowed from other domains of activity and, vice versa, existing debugging analysis and implementation patterns can be applied to other domains, such as to debugging cloud native computing environments[8].

---

[6] Dmitry Vostokov, *Extended Windows Memory Dump Analysis: Using and Writing WinDbg Extensions, Database and Event Stream Processing, Visualization*, Revised 2023 (ISBN-13: 978-191263668)

[7] https://github.com/SeanCline/PyExt

[8] Dmitry Vostokov, "Introducing Methodology and System of Cloud Analysis Patterns (CAPS)," in *Memory Dump Analysis Anthology, Volume 14*, 2021 (ISBN-13: 978-1912636143), pp. 57 – 66

# Gestures

First, I introduce some definitions adapted to debugging from software diagnostics where they were originally introduced[9].

- A debugging action is a user interface action, a command, a technique, a debugging algorithm, and a debugging pattern.

- A space of tools is a collection of physical and virtual (mental, imaginary) tools at some physical or virtual (mathematical) distance from each other.

- A configuration of debugging actions is a directed graph (digraph) in a topological space of tools or a diagram in a category theory[10] sense.

- A debugging gesture is a configuration of debugging actions across the space of tools and time, resulting in a workflow of diagnostic actions.

- A debugging hypergesture is a gesture[11] of debugging gestures, a transformation of one gesture into another, between sets of tools, similar to porting debugging patterns from one platform to another, for example, from Windows to Linux or from one domain to another, for example, from logs to texts (see also the **Lifting** pattern above). You can view debugging hypergestures as debugging gesture patterns.

The "gesture" metaphor stems from the fact that despite recent automation efforts, the debugging process is still manual when it requires substantial domain expertise. We still use various tools, graphical and command line ones (hand movements), and move in cyberspace. So, combining all these physical and virtual movements into some

---

[9] Dmitry Vostokov, *Introducing Diags: Diagnostic Analysis Gestures and Logues, in Theoretical Software Diagnostics: Collected Articles*, Third Edition, 2020 (ISBN-13: 978-1912636334), pp. 131 – 132

[10] https://en.wikipedia.org/wiki/Category_theory

[11] Here, a gesture is informally some action or a sequence of actions, not necessarily related to debugging.

abstract space path is natural. There's also a question of debugging performance (in terms of achieving debugging goals) and repertoire. Debugging gestures also include tool improvisation, data exploration, action experimentation, and aesthetics (coolness, for example). Some gestures can be used to discover further debugging patterns.

# Summary

This chapter briefly introduced debugging usage patterns, the last debugging pattern category. In the next chapter, you will look at some common problems involving Python interfaces with the Windows operating system and use debugging extensions to diagnose and debug them.

# CHAPTER 11

# Case Study: Resource Leaks

In the previous chapter, you looked at debugging usage patterns. In this chapter, you will use the pattern approach to diagnose and debug a common problem type in the Windows operating system: resource leaks.

## Elementary Diagnostics

If you run the handle-leak.py script, you will notice the constant monotone increase of the handles' **Counter Value** (Figure 11-1).

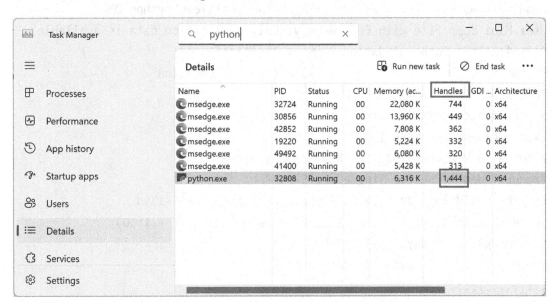

***Figure 11-1.*** *The increasing number of handles for the python.exe process*

© Dmitry Vostokov 2024
D. Vostokov, *Python Debugging for AI, Machine Learning, and Cloud Computing*,
https://doi.org/10.1007/978-1-4842-9745-2_11

You can right-click and save a process memory dump for further offline analysis using the *Create memory dump file* menu option. It is saved in your profile temporary folder (Figure 11-2).

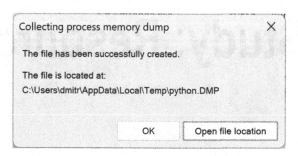

**Figure 11-2.** *Created python.exe process memory dump*

# Debugging Analysis

Now launch the WinDbg debugger and open the saved memory dump[1]:

```
Microsoft (R) Windows Debugger Version 10.0.25921.1001 AMD64
Copyright (c) Microsoft Corporation. All rights reserved.
Loading Dump File [C:\Users\dmitr\AppData\Local\Temp\python.DMP]
User Mini Dump File with Full Memory: Only application data is available
************* Path validation summary **************
Response                       Time (ms)     Location
Deferred                                     srv*
Symbol search path is: srv*
Executable search path is:
Windows 10 Version 22621 MP (8 procs) Free x64
Product: WinNt, suite: SingleUserTS
Edition build lab: 22621.1.amd64fre.ni_release.220506-1250
Debug session time: Sun Sep  3 17:02:35.000 2023 (UTC + 1:00)
System Uptime: 6 days 7:55:29.493
Process Uptime: 0 days 0:09:46.000
................
```

---

[1] When installing Python, you also need to select debug symbol files to have meaningful stack traces.

For analysis of this file, run !analyze -v
ntdll!NtWaitForMultipleObjects+0x14:
00007ff8`6f0ef8a4 ret

You will notice several hundred threads and handles[2]:

```
0:000> ~
.   0  Id: de58.dc60 Suspend: 0 Teb: 000000d1`724e3000 Unfrozen
    1  Id: de58.240c Suspend: 0 Teb: 000000d1`724eb000 Unfrozen
    2  Id: de58.ed7c Suspend: 0 Teb: 000000d1`724ed000 Unfrozen
    3  Id: de58.d404 Suspend: 0 Teb: 000000d1`724ef000 Unfrozen
    4  Id: de58.6f2c Suspend: 0 Teb: 000000d1`724f1000 Unfrozen
    5  Id: de58.3e2c Suspend: 0 Teb: 000000d1`724f3000 Unfrozen
    6  Id: de58.6684 Suspend: 0 Teb: 000000d1`724f5000 Unfrozen
    7  Id: de58.ec74 Suspend: 0 Teb: 000000d1`724f7000 Unfrozen
...
  574  Id: de58.2de8 Suspend: 0 Teb: 000000d1`30adf000 Unfrozen
  575  Id: de58.79f4 Suspend: 0 Teb: 000000d1`30ae1000 Unfrozen
  576  Id: de58.7fdc Suspend: 0 Teb: 000000d1`30ae3000 Unfrozen
  577  Id: de58.4f6c Suspend: 0 Teb: 000000d1`30ae5000 Unfrozen
  578  Id: de58.b1e8 Suspend: 0 Teb: 000000d1`30ae7000 Unfrozen
  579  Id: de58.6658 Suspend: 0 Teb: 000000d1`30ae9000 Unfrozen
  580  Id: de58.4ed8 Suspend: 0 Teb: 000000d1`30aeb000 Unfrozen
  581  Id: de58.5b8c Suspend: 0 Teb: 000000d1`30aed000 Unfrozen
  582  Id: de58.bad8 Suspend: 0 Teb: 000000d1`30aef000 Unfrozen
  583  Id: de58.19cc Suspend: 0 Teb: 000000d1`30af1000 Unfrozen
  584  Id: de58.ba0 Suspend: 0 Teb: 000000d1`30af3000 Unfrozen
  585  Id: de58.cd8c Suspend: 0 Teb: 000000d1`30af5000 Unfrozen
0:000> !handle
Handle 0000000000000004
  Type           File
```

---

[2] I may have a bug in WinDbg !handle extension version I used at the time of this writing or its incompatibility with the possible Windows handle table changes as this version of the debugger doesn't show Thread handle type and there are inconsistencies in the !htrace WinDbg command output.

```
Handle 0000000000000008
   Type            Event
Handle 000000000000000c
   Type            Event
Handle 0000000000000010
   Type            Event
Handle 0000000000000014
   Type            WaitCompletionPacket
Handle 0000000000000018
   Type            IoCompletion
Handle 000000000000001c
   Type            TpWorkerFactory
Handle 0000000000000020
   Type            IRTimer
Handle 0000000000000024
   Type            WaitCompletionPacket
Handle 0000000000000028
   Type            IRTimer
Handle 000000000000002c
...
Handle 0000000000001cc4
   Type            IRTimer
Handle 0000000000001cc8
   Type            IRTimer
Handle 0000000000001ccc
   Type            IRTimer
Handle 0000000000001cd0
   Type            IRTimer
Handle 0000000000001cd4
   Type            Semaphore
Handle 0000000000001cd8
   Type            IRTimer
Handle 0000000000001cdc
   Type            IRTimer
```

**1838 Handles**

| Type | Count |
|------|-------|
| **None** | **616** |
| Event | 7 |
| File | 10 |
| Directory | 2 |
| Mutant | 1 |
| **Semaphore** | **1188** |
| Key | 4 |
| IoCompletion | 2 |
| TpWorkerFactory | 2 |
| ALPC Port | 1 |
| WaitCompletionPacket | 5 |

All threads seem to be Python **Runtime Threads**:

```
0:000> ~585s
ntdll!NtWaitForSingleObject+0x14:
00007ff8`6f0eedd4 ret
0:585> kc
 # Call Site
00 ntdll!NtWaitForSingleObject
01 KERNELBASE!WaitForSingleObjectEx
02 python311!pysleep
03 python311!time_sleep
04 python311!_PyEval_EvalFrameDefault
05 python311!_PyEval_EvalFrame
06 python311!_PyEval_Vector
07 python311!_PyFunction_Vectorcall
08 python311!_PyVectorcall_Call
09 python311!_PyObject_Call
0a python311!PyObject_Call
0b python311!do_call_core
0c python311!_PyEval_EvalFrameDefault
0d python311!_PyEval_EvalFrame
0e python311!_PyEval_Vector
0f python311!_PyFunction_Vectorcall
```

```
10 python311!_PyObject_VectorcallTstate
11 python311!method_vectorcall
12 python311!_PyVectorcall_Call
13 python311!_PyObject_Call
14 python311!thread_run
15 python311!bootstrap
16 ucrtbase!thread_start<unsigned int (__cdecl*)(void *),1>
17 kernel32!BaseThreadInitThunk
18 ntdll!RtlUserThreadStart
```

Based on this diagnostic information, you can confirm it was a thread leak and an additional semaphore leak.

# Debugging Architecture

Since the source code is small (Listing 11-1), let's do **Postmortem** debugging **In Vivo** and **In Situ** (you could also analyze the memory dump **Ex Situ** on another machine) and **In Papyro**. See Chapter 8 for pattern name explanations.

*Listing 11-1.* A Simple Script Illustrating a Handle Leak

```python
# handle-leak.py

import time
import threading

def thread_func():
    foo()

def main():
    threads: list[threading.Thread] = []
    while True:
        thread = threading.Thread(target=thread_func)
        threads.append(thread)
        thread.start()
        time.sleep(1)
```

```
def foo():
    bar()

def bar():
    while True:
        time.sleep(1)

if __name__ == "__main__":
    main()
```

# Debugging Implementation

In complex scenarios, when you have much bigger code potentially involving third-party libraries, you would want to collect the **Usage Trace** for handles to see which thread created them[3]. For postmortem debugging, you need to enable an *application verifier* in 64-bit global flags (`gflags.exe`) on the computer where you are running Python (Figure 11-3).

This `gflags` application is a part of the Debugging Tools for Windows package that you can download and install as a part of WDK or SDK[4]. You can also install these tools on a separate computer and copy `gflags.exe` and `gflagsui.dll` to a computer where you want to set global flags for diagnostics and debugging. If you can't have a GUI on your problem computer (for example, server or Docker environments), then you can set the appropriate registry value, which in my case is the following:

```
Computer\HKEY_LOCAL_MACHINE\SOFTWARE\Microsoft\Windows NT\CurrentVersion\
Image File Execution Options\python.exe
GlobalFlag (REG_DWORD) 0x100 (256)
```

Then, you run your script again and collect the new dump after noticing an increased number of handles. Also, don't forget to clear all flags after you finish troubleshooting and debugging; some flags may affect performance.

---

[3] For Linux, the Valgrind tool can be used for tracing resource handles like file descriptors, for example, https://developers.redhat.com/articles/2023/01/09/how-use-valgrind-track-file-descriptors.

[4] https://learn.microsoft.com/en-gb/windows-hardware/drivers/debugger/debugger-download-tools

When you load the new dump into WinDbg, you can check whether flags were enabled correctly and check handle allocation stack traces:

```
Microsoft (R) Windows Debugger Version 10.0.25921.1001 AMD64
Copyright (c) Microsoft Corporation. All rights reserved.

Loading Dump File [C:\Users\dmitr\AppData\Local\Temp\python (2).DMP]
User Mini Dump File with Full Memory: Only application data is available

************* Path validation summary **************
Response                      Time (ms)     Location
Deferred                                    srv*
Symbol search path is: srv*
Executable search path is:
Windows 10 Version 22621 MP (8 procs) Free x64
Product: WinNt, suite: SingleUserTS
Edition build lab: 22621.1.amd64fre.ni_release.220506-1250
Debug session time: Sun Sep  3 18:18:44.000 2023 (UTC + 1:00)
System Uptime: 6 days 9:11:38.570
Process Uptime: 0 days 0:05:42.000
................
For analysis of this file, run !analyze -v
ntdll!NtWaitForMultipleObjects+0x14:
00007ff8`6f0ef8a4 ret

0:000> !gflag
Current NtGlobalFlag contents: 0x02000100
    vrf - Enable application verifier
    hpa - Place heap allocations at ends of pages

0:000> !htrace
...
Handle = 0x0000000000000ca8 - OPEN
Thread ID = 0x000000000000bed8, Process ID = 0x00000000000083f0

0x00007ff86f0f0624: ntdll!NtCreateThreadEx+0x0000000000000014
0x00007ff86c8aed1f: KERNELBASE!CreateRemoteThreadEx+0x000000000000029f
0x00007ff86c9c708b: KERNELBASE!CreateThread+0x000000000000003b
```

```
0x00007ff83627fbff: verifier!AVrfpCreateThread+0x00000000000000cf
0x00007ff86c52838e: ucrtbase!_beginthreadex+0x000000000000005e
0x00007ff805b99512: python311!PyThread_start_new_thread+0x0000000000000086
0x00007ff805b9925c: python311!thread_PyThread_start_new_
thread+0x0000000000000110
0x00007ff805ba09e6: python311!_PyObject_MakeTpCall+0x0000000000000736
0x00007ff805ba4548: python311!PyObject_Vectorcall+0x00000000000001e8
0x00007ff805ba5da4: python311!_PyEval_EvalFrameDefault+0x0000000000000784
0x00007ff805bc2f73: python311!_PyEval_Vector+0x0000000000000077
...
Handle = 0x0000000000000d20 - OPEN
```

**Thread ID = 0x000000000000bed8**, Process ID = 0x00000000000083f0

```
0x00007ff86f0f0624: ntdll!NtCreateThreadEx+0x0000000000000014
0x00007ff86c8aed1f: KERNELBASE!CreateRemoteThreadEx+0x000000000000029f
0x00007ff86c9c708b: KERNELBASE!CreateThread+0x000000000000003b
0x00007ff83627fbff: verifier!AVrfpCreateThread+0x00000000000000cf
0x00007ff86c52838e: ucrtbase!_beginthreadex+0x000000000000005e
0x00007ff805b99512: python311!PyThread_start_new_thread+0x0000000000000086
0x00007ff805b9925c: python311!thread_PyThread_start_new_
thread+0x0000000000000110
0x00007ff805ba09e6: python311!_PyObject_MakeTpCall+0x0000000000000736
0x00007ff805ba4548: python311!PyObject_Vectorcall+0x00000000000001e8
0x00007ff805ba5da4: python311!_PyEval_EvalFrameDefault+0x0000000000000784
0x00007ff805bc2f73: python311!_PyEval_Vector+0x0000000000000077
...
Handle = 0x0000000000001144 - OPEN
```

**Thread ID = 0x000000000000bed8**, Process ID = 0x00000000000083f0

```
0x00007ff86f0f0624: ntdll!NtCreateThreadEx+0x0000000000000014
0x00007ff86c8aed1f: KERNELBASE!CreateRemoteThreadEx+0x000000000000029f
0x00007ff86c9c708b: KERNELBASE!CreateThread+0x000000000000003b
0x00007ff83627fbff: verifier!AVrfpCreateThread+0x00000000000000cf
0x00007ff86c52838e: ucrtbase!_beginthreadex+0x000000000000005e
0x00007ff805b99512: python311!PyThread_start_new_thread+0x0000000000000086
```

```
0x00007ff805b9925c: python311!thread_PyThread_start_new_
thread+0x0000000000000110
0x00007ff805ba09e6: python311!_PyObject_MakeTpCall+0x0000000000000736
0x00007ff805ba4548: python311!PyObject_Vectorcall+0x00000000000001e8
0x00007ff805ba5da4: python311!_PyEval_EvalFrameDefault+0x0000000000000784
0x00007ff805bc2f73: python311!_PyEval_Vector+0x0000000000000077
...
```

You can see that all additional threads were created by the main python.exe thread:

```
0:000> ~~[bed8]kc
 # Call Site
00 ntdll!NtWaitForMultipleObjects
01 KERNELBASE!WaitForMultipleObjectsEx
02 KERNELBASE!WaitForMultipleObjects
03 verifier!AVrfpWaitForMultipleObjectsCommon
04 verifier!AVrfpKernelbaseWaitForMultipleObjects
05 verifier!AVrfpWaitForMultipleObjectsCommon
06 verifier!AVrfpKernel32WaitForMultipleObjects
07 python311!pysleep
08 python311!time_sleep
09 python311!_PyEval_EvalFrameDefault
0a python311!_PyEval_EvalFrame
0b python311!_PyEval_Vector
0c python311!PyEval_EvalCode
0d python311!run_eval_code_obj
0e python311!run_mod
0f python311!pyrun_file
10 python311!_PyRun_SimpleFileObject
11 python311!_PyRun_AnyFileObject
12 python311!pymain_run_file_obj
13 python311!pymain_run_file
14 python311!pymain_run_python
15 python311!Py_RunMain
16 python311!Py_Main
17 python!invoke_main
```

```
18 python!__scrt_common_main_seh
19 kernel32!BaseThreadInitThunk
1a ntdll!RtlUserThreadStart
```

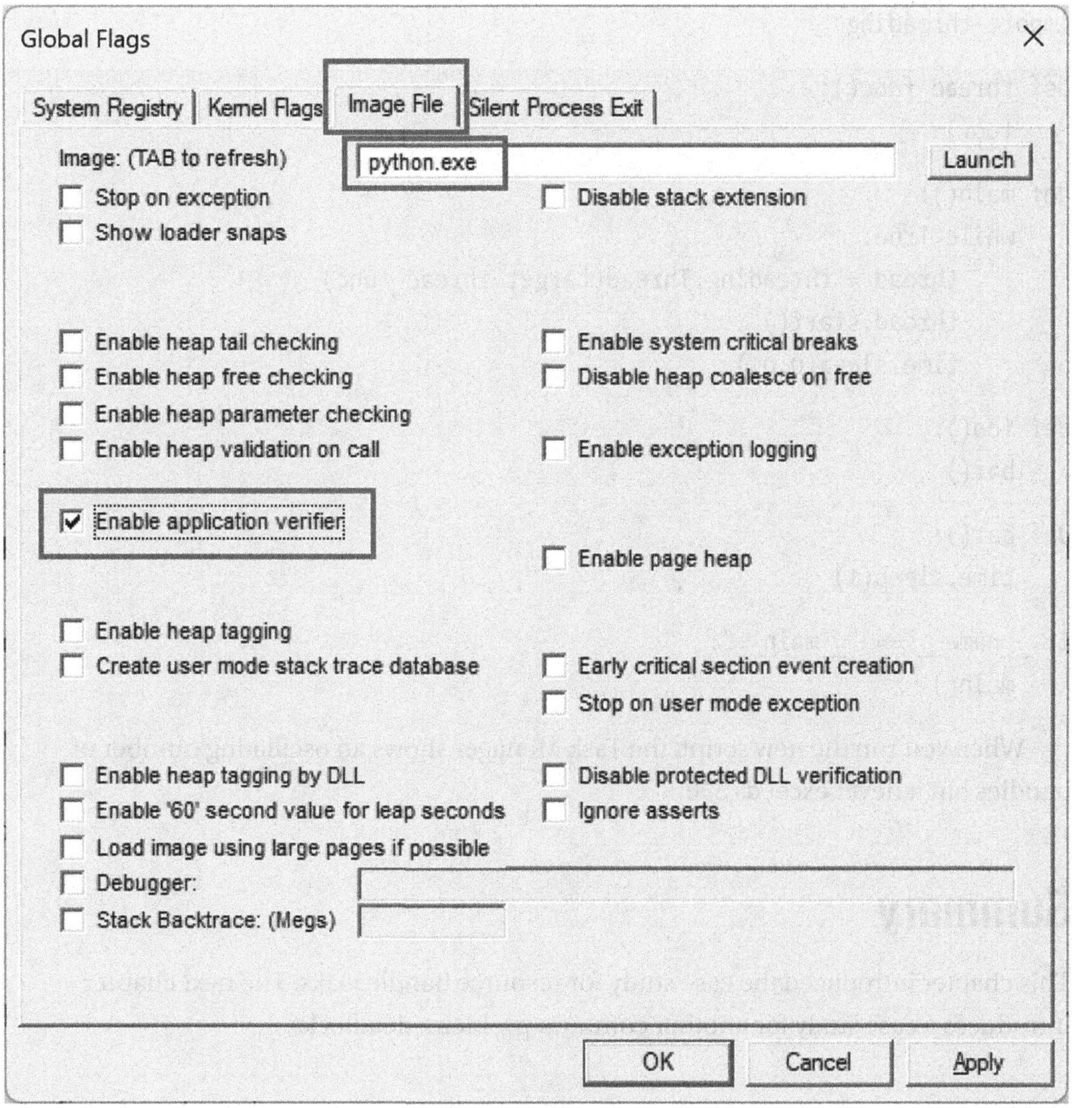

*Figure 11-3.* *Enabling application verifier for python.exe*

You can fix the issue by not keeping thread references and exiting thread function as soon as the thread task is finished, thus allowing thread objects to be garbage collected and the handles closed (Listing 11-2).

***Listing 11-2.*** A Simple Script Illustrating a Fix for the Handle Leak

```python
# handle-leak-fix.py

import time
import threading

def thread_func():
    foo()

def main():
    while True:
        thread = threading.Thread(target=thread_func)
        thread.start()
        time.sleep(0.01)

def foo():
    bar()

def bar():
    time.sleep(1)

if __name__ == "__main__":
    main()
```

When you run the new script, the Task Manager shows an oscillating number of handles but it never exceeds 350.

# Summary

This chapter introduced the case study for resource handle leaks. The next chapter introduces a case study for another common problem: deadlocks.

# Case Study: Deadlock

In the previous chapter, you looked at resource leaks. In this chapter, you will diagnose and debug another common problem type, that of synchronization deadlocks, but you will use Linux as your operation system and GDB as a debugger. Some problem types do not require all debugging pattern layers; only some may be sufficient. For example, with postmortem debugging, sometimes the debugging analysis stage is sufficient.

## Elementary Diagnostics

When you run the deadlock.py script, the process **Hangs** instead of exiting in a few seconds as expected.

```
~/Chapter12$ python3 deadlock.py &
[1] 58
~/Chapter12$ ps
  PID TTY          TIME CMD
    9 pts/0    00:00:00 bash
   58 pts/0    00:00:00 python3
   61 pts/0    00:00:00 ps
```

Let's collect a process core dump for debugging analysis. Alternatively, if GDB is installed, you can attach it to the problem process via gdb -p 58.

```
~/Chapter12$ gcore 58
[New LWP 59]
[New LWP 60]
[Thread debugging using libthread_db enabled]
Using host libthread_db library "/lib/x86_64-linux-gnu/libthread_db.so.1".
```

© Dmitry Vostokov 2024
D. Vostokov, *Python Debugging for AI, Machine Learning, and Cloud Computing*,
https://doi.org/10.1007/978-1-4842-9745-2_12

futex_abstimed_wait_cancelable (private=0, abstime=0x0, expected=0, futex_
word=0x7f7e50000d50) at ../sysdeps/unix/sysv/linux/futex-internal.h:205
205      ../sysdeps/unix/sysv/linux/futex-internal.h: No such file or
directory.
warning: target file /proc/58/cmdline contained unexpected null characters
**Saved corefile core.58**
[Inferior 1 (process 58) detached]

# Debugging Analysis

Open a core dump:

~/Chapter12$ **which python3**
/usr/bin/python3

~/Chapter12$ **gdb -c core.58 -se /usr/bin/python3**
GNU gdb (Debian 8.2.1-2+b3) 8.2.1
Copyright (C) 2018 Free Software Foundation, Inc.
License GPLv3+: GNU GPL version 3 or later <http://gnu.org/licenses/
gpl.html>
This is free software: you are free to change and redistribute it.
There is NO WARRANTY, to the extent permitted by law.
Type "show copying" and "show warranty" for details.
This GDB was configured as "x86_64-linux-gnu".
Type "show configuration" for configuration details.
For bug reporting instructions, please see:
<http://www.gnu.org/software/gdb/bugs/>.
Find the GDB manual and other documentation resources online at:
    <http://www.gnu.org/software/gdb/documentation/>.

For help, type "help".
Type "apropos word" to search for commands related to "word"...
Reading symbols from /usr/bin/python3...Reading symbols from /usr/lib/
debug/.build-id/d7/18c81cb33e7f22039867c673d5d366a849b197.debug...done.
done.

warning: core file may not match specified executable file.
[New LWP 58]
[New LWP 59]
[New LWP 60]
[Thread debugging using libthread_db enabled]
Using host libthread_db library "/lib/x86_64-linux-gnu/libthread_db.so.1".
Core was generated by `python3'.
#0  futex_abstimed_wait_cancelable (private=0, abstime=0x0, expected=0,
futex_word=0x7f7e50000d50)
    at ../sysdeps/unix/sysv/linux/futex-internal.h:205
205     ../sysdeps/unix/sysv/linux/futex-internal.h: No such file or
directory.
[Current thread is 1 (Thread 0x7f7e56c8e740 (LWP 58))]
(gdb)

Now look at the **Managed Stack Trace Collection** using the py-bt command (see Chapter 4's case study on installing the required debugging package scripts).

(gdb) **thread apply all py-bt**
**Thread 3 (Thread 0x7f7e55f42700 (LWP 60)):**
**Traceback (most recent call first):**
  **<built-in method __enter__ of _thread.RLock object at remote**
**0x7f7e56891d80>**
    **File "deadlock.py", line 18, in thread_func2**
      **with cs1:**
    File "/usr/lib/python3.7/threading.py", line 865, in run
      self._target(*self._args, **self._kwargs)
    File "/usr/lib/python3.7/threading.py", line 917, in _bootstrap_inner
      self.run()
    File "/usr/lib/python3.7/threading.py", line 885, in _bootstrap
      self._bootstrap_inner()

**Thread 2 (Thread 0x7f7e56743700 (LWP 59)):**
**Traceback (most recent call first):**
  **<built-in method __enter__ of _thread.RLock object at remote**
  **0x7f7e56891f30>**
  **File "deadlock.py", line 12, in thread_func1**

169

**with cs2:**
```
  File "/usr/lib/python3.7/threading.py", line 865, in run
    self._target(*self._args, **self._kwargs)
  File "/usr/lib/python3.7/threading.py", line 917, in _bootstrap_inner
    self.run()
  File "/usr/lib/python3.7/threading.py", line 885, in _bootstrap
    self._bootstrap_inner()

Thread 1 (Thread 0x7f7e56c8e740 (LWP 58)):
Traceback (most recent call first):
  File "/usr/lib/python3.7/threading.py", line 1048, in _wait_for_
  tstate_lock
    elif lock.acquire(block, timeout):
--Type <RET> for more, q to quit, c to continue without paging--
  File "/usr/lib/python3.7/threading.py", line 1032, in join
    self._wait_for_tstate_lock()
  File "deadlock.py", line 27, in main
    thread1.join()
  File "deadlock.py", line 31, in <module>
    main()
```

Threads #2 and #3 are **Blocked Threads** (introduced in Chapter 4) entering cs2 and cs1 Python objects using the with Python statement. Because these objects are RLocks, you can check their thread ownership without examining the Python script source code. Reentrant locks must keep thread ownership to differentiate between the owner and other threads.

You have the following native information about RLock addresses inside the Python interpreter process from the output above:

```
Thread 3 (Thread 0x7f7e55f42700 (LWP 60)):
 <built-in method __enter__ of _thread.RLock object at remote
0x7f7e56891d80>
Thread 2 (Thread 0x7f7e56743700 (LWP 59)):
  <built-in method __enter__ of _thread.RLock object at remote
0x7f7e56891f30>
```

Let's dump remote addresses' memory and check values from the nearby offsets, the so-called **Pointer Cone**.

```
(gdb) x/4a 0x7f7e56891d80
0x7f7e56891d80: 0x4      0x87d1a0 <RLocktype>
0x7f7e56891d90: 0xe4eec0        0x7f7e56743700
```

```
(gdb) x/4a 0x7f7e56891f30
0x7f7e56891f30: 0x4      0x87d1a0 <RLocktype>
0x7f7e56891f40: 0xe3e110        0x7f7e55f42700
```

From the output above, you see that thread #3 is waiting for the RLock 0x7f7e56891d80 owned by thread #2 (0x7f7e56743700). You also see that thread #2 is waiting for the RLock  0x7f7e56891f30 owned by thread #3 (0x7f7e55f42700). Based on this mutual wait, you hypothesize the **Deadlock**.

# Debugging Architecture

Since the source code is small (Listing 12-1), let's do **Postmortem** debugging **In Situ** and **In Papyro**.

*Listing 12-1.* A Simple Script Illustrating a Deadlock Between Two Threads

```
# deadlock.py

import time
import threading

cs1 = threading.RLock()
cs2 = threading.RLock()

def thread_func1():
    with cs1:
        time.sleep(1)
        with cs2:
            pass
```

```python
def thread_func2():
    with cs2:
        time.sleep(1)
        with cs1:
            pass

def main():
    thread1 = threading.Thread(target=thread_func1)
    thread2 = threading.Thread(target=thread_func2)
    thread1.start()
    thread2.start()

    thread1.join()
    thread2.join()

if __name__ == "__main__":
    main()
```

You immediately see the opposite lock acquisition order in threads. If you correct it (Listing 12-2), the script runs to completion without freezing:

```
~/Chapter12$ python3 deadlock-fix.py
~/Chapter12$
```

***Listing 12-2.*** Code Fix for the Deadlock

```python
def thread_func1():
    with cs1:
        time.sleep(1)
        with cs2:
            pass

def thread_func2():
    with cs1:
        time.sleep(1)
        with cs2:
            pass
```

# Exceptions and Deadlocks

The with is a syntactic sugar for acquire and release methods. If used directly or in the case of third-party and native operating system synchronization objects with the same acquire and release semantics, you may have synchronization problems in the presence of **Handled Exceptions** (Listing 12-3). The exception skips the lock release method, causing a **Deadlock**. In such cases, I recommend augmenting exception handling code with logging to have exception processing visibility.

***Listing 12-3.*** A Script Illustrating a Deadlock Between Two Threads in the Presence of Handled Exceptions

```
# deadlock-exception.py

import time
import threading
import logging

cs1 = threading.RLock()
cs2 = threading.RLock()

def foo (data):
    if data: raise "Exception"

def thread_func1():
    try:
        cs1.acquire()
        foo(1)
        cs1.release()
    except:
        logging.error("Exception logged")

    time.sleep(2)

    cs2.acquire()
    cs2.release()

def thread_func2():
    cs2.acquire()
    cs1.acquire()
```

```python
        time.sleep(3)

        cs1.release()
        cs2.release()
def main():
    thread1 = threading.Thread(target=thread_func1)
    thread2 = threading.Thread(target=thread_func2)
    thread1.start()
    thread2.start()

    thread1.join()
    thread2.join()
if __name__ == "__main__":
    main()
```

# Summary

This chapter introduced a case study for diagnosing and debugging deadlocks using GDB on Linux. The next chapter discusses the challenges of Python debugging in cloud computing.

# CHAPTER 13

# Challenges of Python Debugging in Cloud Computing

In the previous chapter, you looked at deadlock debugging. In this chapter, you will start surveying the challenges of debugging in cloud computing, AI, and machine learning. You will start with cloud computing first, the backbone of modern AI/ML.

Cloud computing is no longer a novelty in today's software development landscape. It provides scalable infrastructure, high availability, and numerous services that aid in building robust applications. When developers transit from local and server-based development environments to cloud platforms, they encounter unique challenges when debugging their Python applications.

## Complex Distributed Systems

The term *complexity* here means intricate but difficult to learn and comprehend. In cloud environments, applications often span multiple services and systems. Debugging becomes challenging when errors arise from these interconnected components. In a monolithic architecture, you might be able to reproduce an issue locally and examine it. However, in a microservices architecture in the cloud, where each service might be written in Python (or other languages) and communicating asynchronously, tracing a problem requires a lot more effort.

When you transition from a local machine or client/server-based setup to the cloud, the first fundamental shift you experience is in the architecture. More often than not, cloud deployments involve distributed systems, especially if we're looking

175

© Dmitry Vostokov 2024
D. Vostokov, *Python Debugging for AI, Machine Learning, and Cloud Computing*,
https://doi.org/10.1007/978-1-4842-9745-2_13

at modern microservices architectures. This transition from localized, monolithic systems to distributed cloud-based ones is no small feat. This paradigm shift comes with an architectural transformation amplifying the manifold debugging challenges. Understanding distributed systems is foundational to debugging in the cloud.

# Granularity of Services

In traditional monolithic and server designs, you often have the luxury of searching for bugs in a singular, unified codebase. With cloud deployments, especially when embracing microservices, your application is fragmented into smaller services. Each could be written in a different language. When a bug manifests, the detective work isn't just about finding the problematic code but also about identifying the service it resides in.

## Service Multiplicity

Distributed systems, particularly those following a microservices architecture, fragment the application into multiple distinct services. Each service has its logic, data flow, and potential pitfalls.

## Localization of Issues

Given the distributed nature, localizing which service is the root cause of a given issue becomes the primary challenge.

# Communication Channels Overhead

Services in distributed systems communicate over the network through HTTP requests, RPCs, or message queues and brokers. Debugging issues arising from these communications, such as failed requests or delayed message deliveries, can be difficult.

## Nature of Communication

Different service communication layers and mechanisms in distributed systems introduce multiple potential failure points.

## Payload Discrepancies

Data being exchanged between services must adhere to expected formats. Mismatches in expected payloads, such as their schemas, can lead to bugs that can be challenging to trace and resolve.

## Latency Concerns

With services hosted potentially across different regions or even continents, network latencies can impact the responsiveness of service-to-service communication.

## Timeout Configurations

Default timeout settings might not always be optimal. Understanding and tweaking these, whether for database connections or HTTP requests, can make a difference.

# Inter-Service Dependencies

One service in a distributed application might depend on several others. If one service fails, it can cause a cascading effect, making understanding the root cause in such scenarios difficult and further complicating the debugging process.

## Service Chaining

In many cloud architectures, especially those serving complex business logic, service calls are often chained, with one service invoking another, and so on. Identifying the weak link when something breaks is very laborious in such chains.

## Service Interactions

Cloud platform Python applications often interact with databases, caches, message queues, and other services. Each interaction is a potential point of failure. Understanding nuances like connection pooling or back-off strategies can be pivotal in debugging.

## Data Consistency

Ensuring robust data consistency in distributed systems is challenging. Eventual consistency models, while mitigating some issues, introduce others like out-of-order data updates or race conditions.

# Layers of Abstraction

Cloud-native platforms layer abstractions over physical resources in their approach to simplify infrastructure management. While beneficial, these abstractions create various debugging challenges.

## Opaque Managed Services

Cloud providers offer various managed services, from databases to machine learning platforms. The services being managed are black boxes to developers. When issues arise within these services, developers have limited insights and controls for debugging.

Managed services come with their own logs. Often more abstract than what one might get from a self-managed service, these logs require interpretation skills. Beyond logs, managed services offer metrics, which can be indispensable in understanding issues. For instance, if a Python application experiences database issues, analyzing metrics like CPU utilization, connection counts, or read/write latencies can offer insights into software behavior.

## Serverless and Function as a Service

Function as a Service (FaaS) solutions abstract away the entire server. Debugging applications in such environments necessitates a paradigm shift, as traditional approaches like using SSH to connect to a compute instance are no longer applicable.

One of the quirks of serverless environments is the concept of cold starts. After being idle, a function's initial invocation can have noticeably higher latencies. Understanding and optimizing for this can be important.

Serverless environments have resource restrictions, like execution time limits or memory constraints. Ensuring that Python applications and pipelines are designed within these confines is vital.

# Container Orchestration Platforms

Platforms like Kubernetes introduce another abstraction layer, where applications run within containers orchestrated by a complex system. Understanding the nuances of these platforms, from pod lifecycles to networking policies, is essential for effective debugging.

Tools like Prometheus[1] can be invaluable in understanding container-specific metrics. Memory leaks or CPU spikes within containers can often be debugged using these insights.

In orchestrated environments, services discover each other dynamically. Ensuring this discovery is seamless and debugging possible issues that may arise there requires understanding underlying systems, for example, etcd[2].

# Continuous Integration/Continuous Deployment

With the rise of DevOps, DevSecOps, and MLOps and the push for faster software delivery, Continuous Integration/Continuous Deployment (CI/CD) pipelines have become ubiquitous in cloud development.

## Pipeline Failures

### Understanding Failures

When a CI/CD pipeline fails due to test failures, linting errors, vulnerability scans, or deployment issues, developers need to look at often verbose logs to pinpoint the root cause.

### Code Analysis Tools

Introducing tools like pylint[3] or flake8[4] can catch potential bugs before they're deployed. However, when these tools flag issues in CI/CD pipelines, developers need to interpret and address them.

---

[1] https://prometheus.io/
[2] https://etcd.io/
[3] www.pylint.org/
[4] https://pypi.org/project/flake8/

## Environment Discrepancies

The *"it works on my machine"* problem can manifest in CI/CD pipelines. Ensuring environment parity using tools like Docker, Rancher Desktop[5], or Colima[6] becomes essential.

## Environment Simulations

The CI/CD pipelines often have stages that simulate different environments, like development, staging, or production. Discrepancies between these simulations and actual environments can lead to unexpected pipeline failures.

# Rollbacks and Versioning

### Identifying Faulty Deployments

In continuous deployment setups, new code gets deployed frequently. If an issue surfaces, identifying the deployment introduced it, especially in Python applications with many dependencies, can be like finding a needle in a haystack.

## Efficient Rollbacks

Once a problem deployment is identified, rolling it back to a stable state is crucial. However, ensuring that rollbacks don't introduce further issues, especially in stateful applications, is challenging.

## Blue-Green Deployments

Techniques like blue-green deployments[7] allow for quick rollbacks. When debugging, developers can switch traffic between versions, isolating issues without affecting all users and dependent services.

---

[5] https://rancherdesktop.io/

[6] https://github.com/abiosoft/colima

[7] https://en.wikipedia.org/wiki/Blue%E2%80%93green_deployment

## Database Migrations

Rollbacks aren't just about code. Rolling back might require a database migration if a deployment introduces a database change. Tools like `alembic`[8] can assist, but understanding their nuances is important.

## Immutable Infrastructure

### State Preservation

Immutable infrastructure principles dictate that the new resources are spun up with desired changes instead of modifying existing ones. While this improves reliability, it can complicate debugging, especially when trying to preserve and understand the state of a problematic resource.

### Resource Proliferation

With frequent deployments and resource replacements, developers can have a proliferation of resources, making it harder to identify which specific instance or version of an application is problematic.

# Diversity of Cloud Service Models

Cloud computing isn't monolithic. It spans diverse service models, each with its own debugging nuances.

## Infrastructure as a Service

### Direct Resource Management

In Infrastructure as a Service (IaaS) setups, developers have direct control over virtual machines and other resources. While this provides flexibility, they are responsible for managing, troubleshooting, and debugging everything from the OS upwards.

---

[8] https://pypi.org/project/alembic/

## Network Complexity

Networking in IaaS environments, with virtual networks, subnets, and security groups, introduces multiple potential points of failure that developers need to be aware of.

# Platform as a Service

## Platform Restrictions

Platform as a Service (PaaS) solutions provide a managed platform for applications. However, developers might not have access to the underlying OS or infrastructure, limiting debugging capabilities.

## Service Limitations

Given that PaaS solutions might offer specific versions of databases, caches, or other services, ensuring compatibility with Python applications and debugging issues arising from service limitations becomes very important.

# Software as a Service

With Software as a Service (SaaS), developers often integrate their applications and pipelines with services that are entirely managed and are essentially black boxes. Debugging issues in such circumstances requires an approach focusing more on API contracts and less on the internals of the service.

# Evolving Cloud Platforms

The cloud landscape isn't static: new services, features, and changes are introduced frequently.

# Adapting to Changes

## Service Evolution

As cloud providers introduce new versions or make incremental changes to their services, developers must ensure their Python applications remain compatible and that any arising bugs are promptly addressed, requiring agility and a deep understanding of the application and the evolving cloud service.

## API Changes

Sometimes cloud providers introduce breaking changes to their published APIs, requiring developers to adapt their Python applications and debug any issues arising from these changes.

## Feature Deprecations

Sometimes cloud providers deprecate features or entire services, and developers must adapt their applications and ensure that any bugs introduced due to such deprecations are quickly resolved.

# Staying Updated

## Continuous Learning

Given the evolution of cloud computing, developers need to continuously learn new services, features, and best practices to debug effectively.

## Community Engagement

Engaging with developer communities on platforms like Stack Overflow and GitHub can provide insights into common bugs or best practices and help with debugging.

## Documentation

Regularly referring to updated documentation of evolving cloud platforms is crucial for effective debugging, ensuring developers aren't chasing ghosts from outdated tools and practices.

# Environment Parity

Local development environments seldom match the cloud environment. Variabilities might range from minor differences in library versions to significant differences in the service infrastructure. As a result, debugging becomes difficult and time-consuming.

## Library and Dependency Disparities

In a Python application or script, the version of libraries and dependencies matters. A version difference between local and cloud environments can lead to a different behavior.

### Version Variabilities

Python applications and scripts are highly reliant on external libraries. Even minor version differences between a local machine and a cloud environment can cause bugs that are hard to reproduce.

### Deprecations and Updates

Libraries evolve, with some features getting deprecated or modified. If the cloud environment updates a library your Python application relies on, it can introduce unforeseen bugs.

## Configuration Differences

Often, configurations for application pipelines might differ based on the environment. Debugging a problem that arises due to misconfigurations can be frustrating and time-consuming.

### Environment-Specific Configs

While configurations like URLs or API keys can differ between environments, more subtle configuration differences can sometimes lead to issues. For instance, a feature or configuration flag might be enabled in one environment but not another.

## Secret Management

In cloud environments, secrets management solutions are often employed. If not properly configured, they can lead to runtime errors in Python applications.

## Underlying Infrastructure Differences

The underlying infrastructure, from the operating system to specific system libraries, can vary between local and cloud environments. Such differences, though subtle, can cause hard-to-diagnose issues.

## Service Variabilities

You might need to integrate with various PaaS offerings in the cloud. These services might not be available or might behave differently in local environments simulated with mock versions or entirely different software stacks. Discrepancies can lead to unexpected behavior when the Python application is deployed to the cloud.

## Limited Visibility

The cloud's ephemeral nature, where instances can be terminated and spun up on-demand, means that logs and other diagnostic data might be lost if not properly managed. Without real-time monitoring and logging solutions integrated, gaining insights into your Python application's behavior can be a difficult task.

## Transient Resources

### Ephemeral Instances

Resources like virtual machines or containers in cloud setups can be short-lived. If an issue arises in a specific instance, developers might find it terminated before they can inspect it, leading to lost debugging information.

## State Replication Challenges

Given the ephemeral nature of these resources, replicating a specific problematic state for debugging becomes a challenge.

# Log Management

Properly configured logging is paramount. With potentially hundreds of instances running simultaneously, aggregating, filtering, and analyzing logs becomes an almost impossible task.

## Volume and Veracity

The volume of logs generated from many instances may be immense. Going through these logs to extract meaningful information may require sophisticated tooling and approaches.

## Centralization Issues

Aggregating logs from myriad sources in a centralized, easily accessible manner is non-trivial. Solutions like ELK (Elasticsearch, Logstash, Kibana) stack[9] or managed services from cloud providers become essential. Developers should also implement structured logging in Python applications, where logs are generated in a structured format (like JSON) that aids in efficient parsing and analysis.

## Contextual Logging

Simple textual log messages may not be sufficient in distributed systems. Adding contextual information, like which service or instance generated a log and under what conditions, becomes very important for effective debugging.

## Correlating Logs

Introducing a trace ID, passed across service calls, can help correlate logs and trace request propagation. Understanding the sequence of events, especially when dealing with asynchronous operations or queued tasks in Python applications, requires correlating logs based on timestamps.

---

[9]`www.elastic.co/elastic-stack/`

# Monitoring and Alerting

Real-time monitoring is crucial in cloud-native environments, but integrating such services into Python applications and setting up meaningful alerts can be a steep learning curve.

## Granular Monitoring

While cloud platforms offer monitoring solutions, developers may need more granular and detailed custom monitoring for a nuanced understanding of Python pipelines. Beyond default metrics, introducing custom metrics in Python applications, like tracking the number of items processed by a background worker or measuring the lag in a processing queue, can offer richer debugging information. Using tools like Grafana[10] to visualize these metrics, perhaps overlaying them with deployment timelines or other events, can help identify patterns or anomalies.

## Alert Fatigue

Too many alerts can lead to alert fatigue, where critical issues might get overlooked amid a sea of notifications. Not all alerts warrant immediate attention. Categorizing them based on severity can ensure critical issues aren't lost in a flow of low-priority notifications. Using monitoring tools with AI and machine learning capabilities can help detect anomalies, reduce false positives, and ensure that developers are only alerted when patterns truly deviate from the norm.

# Latency and Network Issues

Debugging network-related issues in cloud environments is inherently complex. Understanding whether a timeout or a latency issue arises from your Python code, a service you're interfacing with, or a cloud infrastructure problem can be like finding a needle in a haystack. Also, not all failures in Python applications are immediate or obvious. Some manifest slowly, only becoming noticeable over time.

---

[10] https://grafana.com/grafana/

# Network Instabilities

The cloud might promise high availability, but network issues, such as increased latencies or sporadic packet losses, can still affect your Python pipeline performance and functionality.

# Service-to-Service Communication

In a microservices setup, there is intensive service communication. Debugging issues like failed service calls, timeouts, or incorrect data transfers requires understanding the Python code and the underlying network protocols.

# Resource Leaks and Performance

## Slow Degradation

Memory leaks, common in long-running Python applications and pipelines, lead to a slow performance degradation. Detecting and pinpointing such leaks, especially in cloud environments with auto-scaling capabilities, can be challenging.

## Garbage Collection

The Python garbage collector usually handles memory management efficiently. However, certain patterns, like circular references, can introduce leaks. Understanding and debugging these in cloud environments can be challenging.

## Profiling

Tools like cProfile[11] for Python can assist in identifying performance bottlenecks. Integrating these tools with logging and monitoring solutions in cloud environments can provide timely insights.

## Tooling Limitations

While tools may help in local environments, their utility can be limited in cloud setups.

---

[11] https://docs.python.org/3/library/profile.html

# Resource Starvation

## Subtle Indicators

Issues like slowly depleting disk space or gradually increasing CPU usage might not trigger immediate alerts but can lead to major failures over time.

## Throttling

When certain limits are hit, cloud providers often introduce throttling, especially for managed services. Identifying and mitigating such throttling and understanding its root causes becomes essential.

## Auto-scaling

It is important to ensure that resources scale dynamically based on demand, whether they are compute instances or database connections. When auto-scaling rules are misconfigured, it can result in resource starvation.

## External Influences

Sometimes, latent failures are due to external factors. For instance, a service your Python application integrates with may introduce subtle delays, eventually leading to timeouts.

# Concurrency Issues

Cloud platforms, by design, enable applications to operate at a scale with loads of concurrent operations. Issues arising from race conditions or other concurrency-related bugs can be hard to reproduce and debug. Python applications that use `asyncio`[12] for asynchronous operations may offer improved performance. However, with concurrency comes the complexity of debugging.

---

[12] https://docs.python.org/3/library/asyncio.html

# Race Conditions

Race conditions occur when the application's behavior depends on the sequence or timing of uncontrollable events. Reproducing and fixing race conditions in Python, especially in a cloud environment where numerous operations occur simultaneously, can be challenging.

# Deadlocks

Deadlocks happen when two or more operations are waiting for each other to finish, leading the application to freeze. Deadlocks can be a disaster in cloud environments where resources are often shared.

# Security and Confidentiality

For security and confidentiality reasons, debugging activities might be restricted in cloud environments to mitigate possible risks.

## Debugger Access Control Restrictions

Attaching a debugger to a running Python process or opening specific ports might be prohibited in production environments, thus limiting traditional debugging approaches.

## Limited Access

Production environments often have strict access controls. While necessary, this can impede debugging efforts as developers might not have direct access to problematic resources.

## Role-Based Access Controls

Cloud platforms employ role-based access controls (RBACs) to delineate who can access what. Configuring these correctly so debugging is possible without compromising security is not easy.

## Identity and Access Management Policies

Identity and access management (IAM) policies dictate what resources can be accessed. Creating such policies to make debugging feasible without compromising security requires a deep understanding of the underlying IAM model.

## Virtual Private Clouds and Networks

Ensuring that cloud resources are accessible only within secure virtual private clouds (VPCs) or via virtual private networks (VPNs) can limit exposure. Yet, it can also introduce connectivity challenges that developers need to troubleshoot and debug.

# Sensitive Data Exposure

## Logs and Metrics

Logs and metrics, essential for debugging, can inadvertently expose sensitive data. Ensuring that Python applications mask or exclude such data using log sanitization practices while still providing meaningful debugging information can be difficult.

## Data Dumps

Sometimes, developers might need to take data dumps to replicate and debug issues, for example, the **State Dump** debugging analysis pattern introduced in Chapter 4. Ensuring that these dumps don't leak sensitive data and are handled with utmost security is paramount.

## Debug Endpoints

Occasionally, developers introduce debugging endpoints in their services for easier troubleshooting. Ensuring these endpoints are not exposed in production is vital to prevent potential security breaches.

## Data Integrity

Production environments handle real user data. Debugging in such environments demands extreme caution to ensure data integrity and confidentiality.

# Limited Access

Direct access to system parts might be limited in production cloud environments, affecting efforts to replicate and diagnose issues.

# Cost Implications

One can rerun applications in a local environment with minimal costs. However, in the cloud, resources consumed during debugging (like computation, storage, or data transfer) might incur costs. Every action in the cloud often has a price tag, making cost management a legitimate concern during debugging.

## Extended Sessions

Long debugging sessions to replicate issues in environments that mirror production, where resources are continuously utilized, can lead to unexpected costs.

## Resource Provisioning and Deprovisioning

### Temporary Resources

Often, debugging requires spinning up temporary resources such as new instances, databases, or storage. Please make sure these are deprovisioned post-debugging to avoid unnecessary costs.

### Resource Scaling

If, while replicating an issue, developers need to scale resources, this may lead to increased costs. Balancing the need for effective debugging with cost considerations is essential.

## Data Transfer and Storage Fees

Storing logs, especially verbose debug-level logs, can increase storage costs. Transferring data, be it logs, metrics, or data and memory dumps, in and out of the cloud can incur charges, especially if debugging involves moving data across regions or out of the cloud provider's network.

# State Management
## Stateful Services

Cloud-native applications often interact with various stateful services like databases, caches, and storage systems. Reproducing specific states that lead to bugs in your Python application can be challenging, especially considering the vast amount of data and its distributed nature.

## Data Volume

With gigabytes and terabytes of data, isolating and replicating the problematic data subset for debugging can be almost impossible.

# Limited Tooling Compatibility

Debugging tools used in local development might not have cloud counterparts that offer the same features, might not be compatible with specific cloud services, or might lack features in a cloud context, thus forcing developers to adapt to new tools and methodologies with their own learning curves.

# Versioning Issues

Cloud platforms continually evolve, with services getting updates or even being deprecated. A Python application might depend on a specific version of a service, and any changes to that service can introduce unexpected behaviors. Debugging issues arising from such mismatches can be very difficult.

## Deprecations and Changes

If a pipeline relies on a specific cloud service feature now, it may get deprecated or altered later. For Python developers, this means debugging not just the code but also understanding changes in cloud service behavior.

# SDK and Library Updates

Cloud providers regularly update their SDKs. A Python application might break due to changes in these SDKs, adding another dimension to debugging efforts.

# Real-time Debugging and User Experience

Traditional debugging methods, like breakpoints and stopping execution, can be disruptive. When debugging issues in real time, especially in production, there's a need to avoid user and service experience disruption. Finding ways to debug without disruption becomes paramount.

# External Service Dependencies

Many cloud applications in Python depend on external APIs or third-party services. When these services experience outages or bugs, they can manifest as issues in your application. Diagnosing such problems, especially when you lack control or visibility into these services, is challenging.

## Dependency Failures

If a third-party API or service your Python application or pipeline depends on fails or malfunctions, it can manifest as issues within your application. Pinpointing such external causes is challenging, especially when these services are black boxes.

## Rate Limiting and Quotas

External services often impose limits. Hitting these limits through unexpected traffic spikes or misconfiguration can cause application failures that need debugging.

# Asynchronous Operations

Debugging asynchronous code in cloud-based Python pipelines is already complex, and the cloud environment adds additional challenges.

# Flow Tracking

In asynchronous Python applications, the process of tracking the flow of operations that don't follow a linear execution pattern and understanding which operation failed and why can be very challenging.

# Error Propagation

Errors in asynchronous operations might get propagated differently, complicating the debugging process.

# Scaling and Load Challenges

Cloud environments are designed for scalability. But as applications and services scale, new debugging challenges may appear, for example, an issue that only occurs under specific loads or when specific scale thresholds are crossed.

# Load-Based Issues

Some issues in Python applications manifest only under specific loads. Reproducing such scenarios for debugging can be complex, requiring simulated traffic and conditions.

# Resource Contention

As applications scale, they might contend for shared resources. Debugging various issues arising from resource contention, whether database locks or CPU throttling, demands a deep understanding of the application and its use of cloud infrastructure.

# Multi-Tenancy Issues

## Resource Contention

### Isolation

Debugging proper resource isolation can bring challenges, whether at the database level using schemas or at the application level using namespaces.

### Rate Limiting

Introducing rate limiting can prevent a single tenant from monopolizing resources. However, it introduces debugging challenges, especially when legitimate requests are erroneously rate-limited.

## Data Security

Data, at rest or in transit, needs to be encrypted. Ensuring this encryption doesn't introduce performance bottlenecks or other issues may require profiling and debugging. Occasionally, bugs can result in one tenant accessing another's data. Debugging such issues, given the potentiality of such breaches, becomes critical.

# Reliability and Redundancy Issues

## Service Failures

### Failover Mechanisms

Cloud services, while reliable, aren't immune to failures. Implementing failover mechanisms, like replica sets, ensures high availability. However, ensuring that failovers happen seamlessly and debugging related issues can be intricate.

### Backup and Recovery

Ensuring data integrity through backups is vital. However, the backup and recovery process can introduce its own set of bugs.

# Data Durability

## Replication

Data is replicated across zones or regions for durability, especially in distributed cloud environments. Ensuring seamless data replication and consistency across replicas introduces a unique set of debugging challenges.

## Disaster Recovery

Preparing for the worst-case scenario, be it a data center outage or a catastrophic bug, is paramount. Disaster recovery processes must be thoroughly debugged to ensure they restore services to their desired state.

# Summary

While cloud-native platforms offer great benefits for deploying and scaling Python applications, services, and pipelines, they introduce several layers of complexity in the debugging process. A thorough understanding of both Python as a language and the internals of the cloud platform is essential for efficient debugging. Adopting cloud-native debugging tools, security practices, comprehensive monitoring and logging strategies, and striving for a deep understanding of distributed systems are all very important in addressing these challenges effectively.

Another account of some challenges but for a different language can be found in *Practical Debugging at Scale: Cloud Native Debugging in Kubernetes and Production*[13].

The next chapter discusses the challenges of Python debugging in AI and machine learning.

---

[13] Shai Almog, *Practical Debugging at Scale: Cloud Native Debugging in Kubernetes and Production*, Apress, 2023 (ISBN-13: 978-1484290415)

# Challenges of Python Debugging in AI and Machine Learning

In the previous chapter, you looked at the challenges of debugging in the cloud. In this chapter, you will continue surveying debugging challenges in a different domain.

The Python language has established itself as the de facto standard language for many artificial intelligence (AI) and machine learning (ML) projects due to its simplicity and the robustness of its libraries like Pandas, TensorFlow, Keras, PyTorch, and `scikit-learn`. However, debugging Python code in AI/ML introduces unique challenges due to the interplay of data, models, algorithms, and hardware, which aren't as prevalent in conventional software development. This chapter surveys some of these challenges.

## The Nature of Defects in AI/ML

Unlike traditional software, where bugs usually result in crashes, hangs, or wrong outputs, in AI/ML, your model may run smoothly but produce inaccurate or unreliable results with poor accuracy. A defect could be due to faulty data, incorrect algorithm implementation, or an inappropriate choice of model or hyperparameters. The main differences from traditional debugging include

- **Data-driven complexity:** The output isn't just a direct result of your code but depends on your data quality, preprocessing, and inherent patterns.

- **Iterative nature**: Models are improved iteratively. Debugging might often mean improving rather than fixing.

199

© Dmitry Vostokov 2024
D. Vostokov, *Python Debugging for AI, Machine Learning, and Cloud Computing*,
https://doi.org/10.1007/978-1-4842-9745-2_14

## Complexity and Abstraction Layers

The Python language's strength lies in the vast ecosystem of libraries that provide high-level abstractions for complex mathematical operations. While this abstraction aids productivity, debugging through layers of library code can be difficult. There are many optimization techniques used internally by these libraries that can make the flow of execution non-intuitive, making the task of tracing execution to the root cause of the error highly difficult.

## Non-Determinism and Reproducibility

Due to the stochastic nature of many algorithms, particularly those that involve random initialization (like neural networks), running the same code multiple times might yield different results. It makes issues hard to reproduce and even harder to debug. Many frameworks utilize GPU acceleration, which may introduce another layer of non-determinism due to asynchronous operations.

## Large Datasets

AI/ML models often train on large datasets. If there's a bug related to data handling or preprocessing, identifying which specific data point caused the problem is akin to finding a needle in a haystack. Memory issues can arise when handling large datasets, leading to cryptic errors or crashes. These aren't always straightforward to resolve, particularly when the issue is with the way a library handles memory rather than the user's code itself or due to garbage collection issues.

## High-Dimensional Data

ML models, especially deep learning models, often deal with high-dimensional data. Visualizing or understanding data in more than three dimensions is challenging, making it tough to debug issues related to data representation or transformations.

# Long Training Times

Training sophisticated AI/ML models can take a long time, from hours to days or weeks. If a bug manifests after lengthy training, it can be costly regarding time and computational resources, making the debug-edit-retrain loop slow and inefficient.

# Real-Time Operation

Some AI applications need to operate in real time. Debugging these applications without affecting their real-time nature can be a challenge. For example, one can't insert breakpoints in a reinforcement learning agent guiding a self-driving car.

# Model Interpretability

Many modern AI models are often criticized for being *black boxes*. When a model isn't making the expected predictions, it's challenging to understand why; the root cause might be hidden layers deep, making it hard to diagnose. In such cases, debugging often involves more than just code: it's about understanding the model's logic, which can be non-intuitive.

# Hardware Challenges

With the rise of GPU-accelerated training and inference, developers often run into hardware-specific issues. These issues can include compatibility issues, driver errors, or hardware-specific bugs. Debugging on GPUs can be harder than on CPUs due to limited support for traditional debugging tools and the parallel nature of GPU operations. Modern ML pipelines often employ clusters of machines. While enabling scalability, distributed computing introduces challenges like synchronization issues, data distribution problems, and network-related bugs.

# Version Compatibility and Dependency Hell

AI/ML libraries are rapidly evolving. When run with a newer library version, code that worked perfectly with the previous one might produce errors (or worse, subtly incorrect results). Projects often rely on multiple libraries, which might have inter-dependencies leading to hard-to-trace errors.

# Data Defects

Data forms the bedrock of any AI/ML project. Given its importance, a majority of debugging often starts here.

## Inconsistent and Noisy Data

Ensure that data is consistently labeled. Mismatched labels or noisy data can drastically reduce model accuracy. Data analysis libraries like `pandas` offer functionalities like `describe()` or `info()`, which give an initial insight into the data's nature. Visualization libraries such as `matplotlib` and `seaborn` can assist in spotting outliers or anomalies.

## Data Leakage

This problem occurs when the model accidentally gets access to the target variables during training. You need to be sure that validation and test data are kept completely separate from the training set. For example, the `scikit-learn` library has the functionality to ensure proper data separation, `train_test_split`. Regular data audits and an understanding of data sources are essential.

## Imbalanced Data

If one class of data dominates (the so-called class imbalance), the model might predict that class by default and skew model performance. Various techniques like oversampling the minority class, undersampling the majority class, or using synthetic data can help.

## Data Quality

Missing values, outliers, and duplicate entries can introduce bugs. Again, libraries like `pandas` or equivalent help inspect and preprocess your data.

## Feature Engineering Flaws

Incorrectly engineered features can mislead models. Using feature importance techniques can help in selecting and refining features.

# Algorithmic and Model-Specific Defects

Once data integrity is established, one should look at the model itself.

## Gradients, Backpropagation, and Automatic Differentiation

Neural networks rely on gradient-based optimization. Issues like vanishing (too small) or exploding (too large) gradients can cause training problems without any clear error message. Here, techniques like gradient clipping, batch normalization, appropriate weight initialization, and using different activation functions can help. Automatic differentiation libraries can sometimes produce incorrect gradients due to internal bugs or numerical instability. Debugging requires a deep understanding of these concepts for manual verification.

## Hyperparameter Tuning

Incorrectly chosen hyperparameters might make a model appear to be malfunctioning when, in fact, it's operating as intended but with suboptimal settings. Distinguishing between genuine bugs and poor hyperparameter choices can be a real challenge requiring experience and, often, exhaustive experimentation. Algorithms to search for the best hyperparameters can have bugs, adding another layer of complexity.

## Overfitting and Underfitting

These are classic challenges in machine learning. Overfitting can make a model perform exceptionally well on training data but poorly on unseen data, appearing as a bug. But often, the issue lies in the model architecture, data, or training process. Debugging overfitting or underfitting (behaving poorly on training data) often means revisiting the model architecture, data splitting strategies, or regularization techniques, adding more complexity to the diagnostics and debugging.

# Algorithm Choice

Not every algorithm suits every problem type. Choosing the correct algorithm based on data size, feature types, and the nature of the problem is very important.

# Deep Learning Defects

Deep learning, given its complexity, brings its own set of challenges.

## Activation and Loss Choices

The choice of activation functions in the layers and the loss function can influence model training.

## Learning Rate

A too-high learning rate can cause the model to diverge, while a too-low rate can cause slow convergence. Learning rate schedules or adaptive learning rates can be employed.

# Implementation Defects

## Tensor Shapes

Mismatched tensor shapes can lead to runtime errors or unexpected behaviors.

## Hardware Limitations and Memory

Large dataset training, models, and deep networks require significant memory, and out-of-memory errors are common. There's a need for continuous GPU/CPU usage and memory monitoring. Tools like `nvidia-smi`[1] can be useful here.

---

[1] `https://developer.nvidia.com/nvidia-system-management-interface`

# Custom Code

If you have written custom loss functions, layers, or evaluation metrics, ensure they are implemented correctly by always testing individual components before integrating them.

# Performance Bottlenecks

Profiling the Python code to identify sections that might be slowing down the process may be necessary. cProfile and line_profiler[2] can pinpoint bottlenecks.

# Testing and Validation

## Unit Testing

Use unittest[3] or pytest[4] for writing unit tests. In the AI/ML context, they can be used for testing data preprocessing functions or custom model components.

## Model Validation

Always split your data into training, validation, and test sets. Monitor your model's performance on the validation set during training to avoid overfitting.

## Cross-Validation

A more robust method than a simple train/test split, cross-validation can provide better insights into how your model might perform on unseen data and a more holistic view of the model performance.

---

[2] https://github.com/pyutils/line_profiler
[3] https://docs.python.org/3/library/unittest.html
[4] https://docs.pytest.org/

# Metrics Monitoring

It's essential to keep track of metrics like accuracy, precision, recall, F1 score, or AUC-ROC (for classification) and MSE or RMSE (for regression). Watching these metrics over time gives an understanding of where the model might falter.

# Visualization for Debugging

Always visualize the data and model outputs.

# TensorBoard

TensorBoard, a tool for TensorFlow[5] and PyTorch[6], helps visualize model architectures, monitor training progress, explore high-dimensional data, and monitor training processes.

# Matplotlib and Seaborn

Plotting your data can reveal patterns or anomalies that are otherwise hidden.

# Model Interpretability

Tools like SHAP[7] or LIME[8] can assist in understanding how models make decisions, thus providing insights into potential model biases or errors.

# Logging and Monitoring
## Checkpoints

Model checkpoints should be saved at regular intervals.

---

[5] www.tensorflow.org/tensorboard
[6] https://pytorch.org/tutorials/recipes/recipes/tensorboard_with_pytorch.html
[7] https://github.com/shap/shap
[8] https://github.com/marcotcr/lime

# Logging

The Python `logging` module can record training progress, encountered anomalies, and performance metrics.

# Alerts

Setting up alerts for potential bugs or completion notifications for longer training sessions on cloud platforms can be very useful.

# Error Tracking Platforms

Tools like Sentry[9] can capture runtime errors and exceptions, providing real-time insights into issues.

# Collaborative Debugging

The AI/ML community is a great resource for this rapidly evolving field.

# Forums and Communities

Websites like Stack Overflow, Reddit's r/MachineLearning, or AI/ML-focused forums like fast.ai are rich sources of collaborative knowledge. Someone might have already faced (and solved) your current problem.

# Peer Review

Having another set of eyes on your code can help catch mistakes you might have overlooked. Having peer data scientists or developers review your code can highlight overlooked areas.

---

[9] `https://sentry.io/`

# Documentation, Continuous Learning, and Updates

## Maintaining Documentation

Given the iterative nature of ML projects, maintaining comprehensive documentation helps keep track of changes, decisions, and encountered issues.

## Library Updates

AI/ML libraries are continually updated. Use stable versions and stay updated with the latest changes or bug fixes.

## Continuous Learning

AI/ML research introduces new techniques and tools regularly, so you need to keep an eye on arXiv, conferences, and blogs to stay updated on new techniques and tools that may be useful for debugging.

## Case Study

Suppose you have a dataset of product sales that includes columns for the product name, date of sale, and price. Over time, you notice that the data is inconsistent (different names for the same product) and noisy (for example, unrealistically high sale prices).

The first step is to do data exploration (Listing 14-1). You get the following output:

```
\Chapter14> python .\inconsistent-noisy-data-exploration.py
    product        date   price
0   productA  2023-08-01    10.0
1   ProductA  2023-08-02    11.0
2   productB  2023-08-03     6.0
3   ProDuctB  2023-08-03     5.9
4   productC  2023-08-04     9.0
5   productA  2023-08-05    10.0
6   productA  2023-08-06  1000.0
7   productC  2023-08-07     9.5
```

***Listing 14-1.*** A Simple Script Illustrating Data Exploration

```
# inconsistent-noisy-data-exploration.py

import pandas as pd

dataset = {
    'product': ['productA', 'ProductA', 'productB', 'ProDuctB', 'productC',
    'productA', 'productA', 'productC'],
    'date': ['2023-08-01', '2023-08-02', '2023-08-03', '2023-08-03',
    '2023-08-04', '2023-08-05', '2023-08-06', '2023-08-07'],
    'price': [10, 11, 6, 5.9, 9, 10, 1000, 9.5]
}

df = pd.DataFrame(dataset)
print(df)
```

Visualizing data can often help identify outliers or noisy data points. In this case study, a box plot helps identify unusual prices (Listing 14-2 and Figure 14-1).

***Listing 14-2.*** A Simple Script Illustrating Data Visualization

```
# inconsistent-noisy-data-visualization.py

import pandas as pd
import matplotlib.pyplot as plt

dataset = {
    'product': ['productA', 'ProductA', 'productB', 'ProDuctB', 'productC',
    'productA', 'productA', 'productC'],
    'date': ['2023-08-01', '2023-08-02', '2023-08-03', '2023-08-03',
    '2023-08-04', '2023-08-05', '2023-08-06', '2023-08-07'],
    'price': [10, 11, 6, 5.9, 9, 10, 1000, 9.5]
}

df = pd.DataFrame(dataset)
plt.boxplot(df['price'])
plt.title('Box plot of prices')
plt.show()
```

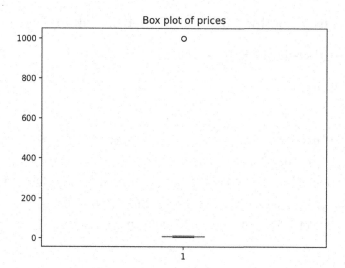

***Figure 14-1.*** *A sample box plot of prices*

From the initial data exploration and visualization, you can identify two issues:

1.   Inconsistent product names due to different capitalizations

2.   Noisy prices (the price of 1000.0 for productA is likely an error)

You fix these issues by normalizing product names and filtering out noisy prices. Then you repeat data exploration and visualization again (Listing 14-3 and Figure 14-1):

```
\Chapter14> python .\inconsistent-noisy-data-cleaning.py
     product        date  price
0   producta  2023-08-01   10.0
1   producta  2023-08-02   11.0
2   productb  2023-08-03    6.0
3   productb  2023-08-03    5.9
4   productc  2023-08-04    9.0
5   producta  2023-08-05   10.0
7   productc  2023-08-07    9.5
```

***Listing 14-3.*** A Simple Script Illustrating Data Cleaning

```python
# inconsistent-noisy-data-cleaning.py

import pandas as pd
import matplotlib.pyplot as plt

dataset = {
    'product': ['productA', 'ProductA', 'productB', 'ProDuctB', 'productC',
    'productA', 'productA', 'productC'],
    'date': ['2023-08-01', '2023-08-02', '2023-08-03', '2023-08-03',
    '2023-08-04', '2023-08-05', '2023-08-06', '2023-08-07'],
    'price': [10, 11, 6, 5.9, 9, 10, 1000, 9.5]
}

df = pd.DataFrame(dataset)

df['product'] = df['product'].str.lower()

mean_price = df['price'].mean()
std_price = df['price'].std()

df = df[(df['price'] < mean_price + std_price) & (df['price'] > mean_
price - std_price)]

print(df)
plt.boxplot(df['price'])
plt.title('Box plot of prices')
plt.show()
```

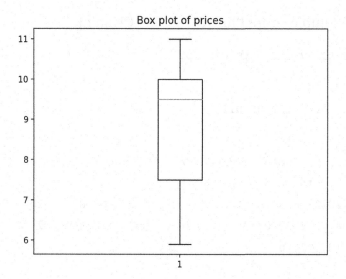

***Figure 14-2.*** *A sample box plot of prices after data cleaning*

# Summary

In conclusion, debugging in AI/ML is challenging due to the intersection of traditional programming and software internals, domain knowledge, complexities of algorithms and the data they process, and the large number of Python libraries and their abstraction. While the tools and practices for debugging in AI/ML continue to evolve, a deep understanding of the underlying mathematics and Python internals remain crucial for effective debugging problem solving.

The next chapter discusses what AI/ML can bring to Python debugging.

# CHAPTER 15

# What AI and Machine Learning Can Do for Python Debugging

In the previous chapter, you looked at debugging AI and machine learning systems written in Python. In this chapter, you will survey the possibilities of using the rapidly evolving AI/ML to aid, enhance, and even revolutionize the traditional manual Python debugging process.

## Automated Error Detection

Before a developer realizes there might be an error or vulnerability in code due to incorrect logic or external dependency, AI/ML systems can proactively scan and highlight potential areas of concern. By analyzing vast amounts of error patterns and fixes from various codebases, machine learning models can be trained to recognize and predict problematic areas in new, unseen code and then detect anomalies and highlight potential errors. The application of ML can make this prediction more accurate over time, reducing false positives, and can lead to drastic reductions in debugging time.

## Intelligent Code Fix Suggestions

Just as some platforms and IDEs suggest word completions based on user typing, syntax, and standard libraries, AI can recommend code snippets and corrections beyond autocompletion. Once an error is detected, AI can list potential fixes that might not have been immediately evident and ranked by relevance and likelihood. It can understand the

© Dmitry Vostokov 2024
D. Vostokov, *Python Debugging for AI, Machine Learning, and Cloud Computing,*
https://doi.org/10.1007/978-1-4842-9745-2_15

programmer's context and intent and use that understanding to provide a starting point of investigation and further meaningful suggestions, reducing trial-and-error and the cognitive load and streamlining the debugging process.

## Interaction Through Natural Language Queries

Traditionally, debugging tools and IDEs have relied on specific command-line inputs or GUI interactions (see the **Gestures** debugging usage pattern introduced in Chapter 10). However, these might not always be intuitive for all developers, especially beginners. Imagine asking your IDE, "Is there a deadlock here?" and getting a direct, relevant answer. Such interactions are becoming possible with natural language processing (NLP). By combining NLP with debugging tools, developers can interact with their development environment more intuitively, making pinpointing potential issues easier. As these AI systems interact with diverse queries over time, their comprehension and responsiveness are only enhanced.

## Visual Debugging Insights

ML systems can enhance visual debugging tools and automatically construct visual representations, flowcharts, and diagrams to aid the understanding of complex systems. Instead of manually tracing through code, visual representations can show the data flow and highlight potential error points. By analyzing the execution and data flow, dependencies, and interactions, ML can predict which parts of the code are most likely to fail and visually represent these predictions, highlight potential bottlenecks, areas of high complexity, or segments of the code that have a higher probability of errors based on historical data, thus assisting developers in focusing their debugging efforts. These insights are particularly valuable for large projects or when working with unfamiliar codebases, offering a holistic bird's-eye view while pinpointing areas of concern.

## Diagnostics and Anomaly Detection

For larger codebases or projects, especially those involving real-time data stream processing or complex algorithms, it's hard to manually detect outliers or anomalies in data flow, processing, or output. Machine learning models, especially those designed for anomaly detection, can automatically flag unexpected or unusual behavior patterns, thus making them particularly useful in scenarios like performance debugging where the issues aren't straightforward syntax or logic errors. Such AI-driven anomaly detection ensures robustness in applications, which is particularly critical for sectors like finance, healthcare, or aerospace, where anomalies can have severe consequences. As these models become increasingly sophisticated, they detect and predict potential anomalies, facilitating preemptive debugging.

# Augmenting Code Reviews

Code reviews are essential to ensure and maintain code quality. However, humans can overlook or miss potential problems. AI-powered tools can be used alongside human reviewers as a second set of eyes, scanning through code changes and highlighting areas that might violate best practices, potential performance issues, or even segments that historically have been problematic. These tools can be trained on coding best practices, ensuring a more comprehensive review process. This also provides an opportunity for continuous learning for developers, who receive immediate feedback on their code. This also saves time for other human reviewers, and that time can be better spent elsewhere. One of the examples is VS Code with integrated CoPilot.

# Historical Information Analysis and Prognostics

ML models can predict potential future bugs by analyzing the historical data of a project, including comments, development decisions, previously encountered bugs and their fixes, and associated metadata. Predictive debugging can preemptively highlight areas of the code that might become problematic in the future based on patterns observed in the past and radically reduce the long-term technical debt of a project. Developers can ensure cleaner, more stable code by foreseeing and addressing issues early on. As these models are exposed to diverse historical patterns across various projects, their predictive accuracy amplifies.

# Adaptive Learning and Personalized Debugging Experience

Every developer has a unique coding style. Over time, AI-enhanced debugging tools can learn individual preferences, common mistakes, and preferred solutions a particular developer makes. By doing so, it can offer more tailored suggestions and solutions, making the debugging process smoother and more efficient. Personalized debugging speeds up the error-resolution process and offers a more enjoyable and less frustrating development experience, thus leading to highly personalized IDEs that adapt to the individual developer's needs and preferences.

# Test Suite Integration and Optimization

Automated testing is critical to modern software development to ensure functionality and prevent regressions. AI can be integrated into testing suites to detect failures and suggest potential code modifications that would rectify the issues. Additionally, by analyzing patterns of test failures, machine learning can optimize the order of test execution, focusing on those most likely to fail first. Integrating AI with test suites can lead to more resilient codebases. Over time, it can also help generate tests, ensuring comprehensive coverage and robustness.

# Enhanced Documentation and Resource Suggestions

Upon encountering an error, AI/ML can instantly suggest relevant documentation, forums, community discussions, video tutorials, or other resources that might help resolve the issue, thus reducing developers' time searching for solutions online. By offering these tailored resources, developers can find solutions faster and expand their understanding. Over time, as such systems become better at understanding the developer's context, the quality and relevance of these suggestions grows, making for a seamless debugging experience.

# Problem Modeling

Another aspect often overlooked in debugging processes is modeling complex software issues with smaller code to understand specific real problems better and provide hypotheses. AI/ML systems can help generate code specific for operating systems and platforms, embedded in particular frameworks and surrounded by required libraries. Generation of boilerplate code for such models and providing alternative models can greatly reduce valuable time.

# Generative Debugging Strategy

This book introduced the pattern-oriented debugging process with debugging stages and debugging pattern languages, including hundreds of pattern names, especially for diagnostic analysis of memory dumps, traces, and logs. Over time, AI/ML systems learn to generate debugging strategies using debugging patterns based on problem descriptions and help communicate debugging results.

# Help with In Papyro Debugging

Large language models (LLM) trained on code repositories can help with the **In Papyro** debugging architecture pattern introduced in Chapter 9. For example, if you feed the Python code example from Chapter 12 to the GPT-4 model and ask the question *What's wrong with this Python code?*, you will get the response in Listing 15-1 including the code fix.

***Listing 15-1.*** A Response from the GPT-4 Model to a Question About Code

```
The Python code you provided demonstrates a classic case of a deadlock.
Here's a step-by-step explanation of the potential deadlock:

thread_func1 acquires cs1 and then goes to sleep for 1 second.
Almost simultaneously, thread_func2 acquires cs2 and then also goes to
sleep for 1 second.
After waking up, thread_func1 will attempt to acquire cs2, but it is
currently held by thread_func2. So, thread_func1 will block, waiting for
cs2 to be released.
```

At the same time, thread_func2 will wake up and attempt to acquire cs1, but cs1 is already acquired by thread_func1. So, thread_func2 will block as well.

At this point, both threads are blocked, waiting for the other to release the lock. This situation will never resolve itself, and it is what we call a deadlock.

To avoid this, you must ensure that all threads acquire the resources (in this case, locks) in the same order. For example, always acquire cs1 before cs2 in all threads:

```python
def thread_func1():
    with cs1:
        time.sleep(1)
        with cs2:
            pass

def thread_func2():
    with cs1:
        with cs2:
            time.sleep(1)
            pass
```

This way, even if one thread has acquired cs1, the second thread will block on cs1 and won't proceed to hold cs2, thereby preventing the deadlock situation.

# Summary

Python, one of the most popular programming languages, has a vast ecosystem of tools and libraries. Introducing AI/ML into the debugging workflow can revolutionize how developers approach problem-solving. While the human touch in debugging (intuition and experience) is always irreplaceable, AI can act as a valuable assistant, reducing the time spent on error resolution and enhancing code quality. As these technologies gradually improve, we can expect a future where debugging is more of a collaborative effort between human and machine, making software development more efficient and debugging more enjoyable.

# CHAPTER 16

# The List of Debugging Patterns

This chapter lists debugging pattern names to facilitate easy recall.

## Elementary Diagnostics Patterns

Introduced in Chapter 3:

- Use-case deviation

- Crash

- Hang

- Counter Value

- Error Message

## Debugging Analysis Patterns

Introduced in Chapter 4:

- Paratext

  - State Dump

  - Counter Value

  - Stack trace patterns

  - Stack Trace

© Dmitry Vostokov 2024
D. Vostokov, *Python Debugging for AI, Machine Learning, and Cloud Computing*,
https://doi.org/10.1007/978-1-4842-9745-2_16

- Runtime Thread

- Managed Stack Trace

- Source Stack Trace

- Stack Trace Collection

- Stack Trace Set

- Exception patterns

  - Managed Code Exception

  - Nested Exception

  - Exception Stack Trace

  - Software Exception

- Module patterns

  - Module Collection

  - Not My Version

  - Exception Module

  - Origin Module

- Thread patterns

  - Spiking Thread

  - Active Thread

  - Blocked Thread

  - Blocking Module

- Synchronization patterns

  - Wait Chain

  - Deadlock

  - Livelock

- Memory consumption patterns

  - Memory Leak

  - Handle Leak

# Debugging Architecture Patterns

Introduced in Chapter 8:

- Where?

  - In Papyro

  - In Vivo

  - In Vitro

  - In Silico

  - In Situ

  - Ex Situ

- When?

  - Live

  - JIT

  - Postmortem

- What?

  - Code

  - Data

  - Interaction

- How?

  - Software Narrative

  - Software State

# Debugging Design Patterns

Introduced in Chapter 9:

- Mutation

- Replacement

- Isolation

- Try-Catch

# Debugging Implementation Patterns

Introduced in Chapter 5:

- Break-in

- Code Breakpoint

- Code Trace

- Scope

- Variable Value

- Type Structure

- Breakpoint Action

- Usage Trace

# Debugging Usage Patterns

Introduced in Chapter 10:

- Exact Sequence

- Scripting

- Debugger Extension

- Abstract Command

- Space Translation

- Lifting

- Gestures

# Debugging Presentation Patterns

Introduced in Chapter 7:

- REPL

- Breakpoint Toolbar

- Action Toolbar

- State Dashboard

# Index

## A

Abstract command, 152
Abstraction layers, 200
    container orchestration, 179
    managed services, 178
    serverless and FaaS, 178
Action toolbar, 125
Active thread, 48
Adaptive learning, 216
Alert fatigue, 187
Alerts, 207
Algorithm choice, 204
Artificial intelligence (AI)
    anomaly detection, 214
    automated error detection, 213
    case study, 208
    code reviews, 215
    complexity and abstraction layers, 200
    datasets, 200
    debugging strategy, 217
    dependency hell, 201
    diagnostics, 214
    hardware challenges, 201
    high-dimensional data, 200
    historical data, 215
    model interpretability, 201
    natural language, 214
    non-determinism and
        reproducibility, 200
    problem modeling, 217
    real-time operation, 201

    training times, 201
    version compatibility, 201
    visual debugging, 214
Asynchronous Operations, 194, 195
Automated error detection, 213
Automated testing, 216
Automatic differentiation, 203
Auto-scaling, 189

## B

Backpropagation, 203
Backtrace, 6–12, 31, 38, 57, 58, 105
Blocked thread, 48
Blocking module, 48
Blue-green deployments, 180
Break-ins, 66–70, 102, 118, 147
Breakpoint actions, 78–81, 104, 147
Breakpoint toolbar, 125

## C

Checkpoints, 206
Class imbalance, 202
Cloud computing
    communication channels, 176, 177
    cost implications
        data transfer and storage fees, 192
        de-provisioning, 192
        extended sessions, 192
        resource provisioning, 192
    external service dependencies